THE CHEMISTRY OF CONJUGATED
CYCLIC COMPOUNDS

THE CHEMISTRY OF CONJUGATED CYCLIC COMPOUNDS

To Be Or Not To Be
Like Benzene?

Douglas Lloyd

Department of Chemistry, University of St Andrews, St Andrews, Fife, Scotland

JOHN WILEY & SONS

Chichester · New York · Brisbane · Toronto · Singapore

Other Wiley Editorial Offices

John Wiley & Sons, Inc., 605 Third Avenue,
New York, NY 10158-0012, USA

Jacaranda Wiley Ltd, G.P.O. Box 859, Brisbane,
Queensland 4001, Australia

John Wiley & Sons (Canada) Ltd, 22 Worcester Road,
Rexdale, Ontario M9W 1L1, Canada

John Wiley & Sons (SEA) Pte Ltd, 37 Jalan Pemimpin 05–04,
Block B, Union Industrial Building, Singapore 2057

Library of Congress Cataloging-in-Publication Data:

Lloyd, Douglas.
    The chemistry of cyclic conjugated compounds : to be or not to be
    like benzene?/Douglas Lloyd.
        p.   cm.
    Includes bibliographical references.
    ISBN 0 471 91721 4
    1. Cyclic compounds. I. Title.
    QD331.L6   1990   89-22416
    547′.5—dc20   CIP

British Library Cataloguing in Publication Data:

Lloyd, Douglas
    The chemistry of cyclic conjugated compounds : to be
    or not to be like benzene?
    1. Cyclic organic compounds
    I. Title
    547′.5

    ISBN 0 471 91721 4

Phototypesetting by Thomson Press (India) Limited, New Delhi
Printed and bound in Great Britain by Biddles Ltd., Guildford.

Contents

Preface

Professor Wilson Baker first introduced me to the fascinating field of what were then generally called 'Non-benzenoid Aromatic Compounds', and ever since I have continued to be delighted by the intriguing chemistry which these compounds offer, covering so many aspects of chemistry, molecular and electronic structures, steric influences, spectroscopy, interplay of calculation and experiment, and, not least, by the sheer intellectual and aesthetic pleasure they provide. I am deeply indebted to Wilson Baker for providing me with this so-satisfying interest.

Out of this interest have come two books, *Carbocyclic Non-benzenoid Aromatic Compounds* (Elsevier, 1966) and *Non-benzenoid Conjugated Carbocyclic Compounds* (Elsevier, 1984). A comparison between them shows the enormous growth in the subject, both practical and theoretical, in the intervening years. Both these books were aimed at specialist readers, and tried to give a reasonably comprehensive and detailed coverage of their topic, with full references, subject, of course, to the constraints of space and of not making them too voluminous.

When it was suggested to me by John Wiley that a student text on the same subject, written in a similar spirit, would be very worthwhile, I happily accepted this invitation. Despite the interest and importance of the compounds discussed in this book, they are largely ignored or are referred to only briefly in the standard textbooks on organic chemistry. It is hoped that this book may make amends for that omission.

The present text is not just a condensed version of the others. It has been newly written with the different needs of students in mind, and aims to be a teaching text rather than a review of the field. Therefore I have changed the order in which I have developed the subject, to provide what I hope is a logical presentation of the facts and ideas, leading the reader through the subject from his or her knowledge of benzene, obtained from standard textbooks and lectures on organic chemistry, to the many facets of the chemistry of cyclic conjugated compounds. The text is aimed at Honours BSc students and at graduate readers who require a more general and introductory approach to the subject than that provided by my earlier books. In general, references are not provided; if more detailed information and references to the original literature are required they may be obtained, except for the most recent work, through my 1984 book. At the ends of the chapters a few general references are given. A number of these are to older articles which give more information about the early development of different topics than appears in this text. Others refer to useful reviews of special topics.

As a glance at the Contents will show, I start from benzene and then go on to compare other cyclic polyenes with it, following this with a discussion of fully conjugated ionic species and finally presenting some chapters dealing with bicyclic and polycyclic compounds. There is a very brief final chapter mentioning the subject of aromatic transition states in pericyclic reactions, but this is very much a tailpiece just to show the relationship of this to the main subject matter of the text. Further detail about pericyclic reactions can be found in a number of excellent textbooks on that topic.

Although most consideration is given to carbocyclic compounds, heterocyclic compounds are not ignored. Some discussion of heterocyclic analogues is found in most chapters, where they are compared with the corresponding carbocyclic compounds. In general this seems to be a better way of approaching heterocyclic compounds, to show the features they have in common with carbocyclic compounds rather than to separate them off as somehow different. With the exception of ferrocene, which played such an important role in the development of the chemistry of cyclopentadiene derivatives that it cannot be excluded, no attention is paid to organometallic compounds. This is partly because lines have to be drawn somewhere about what is and what is not to be included in a text to keep it within set bounds, and this is clearly a defined place where it can be done. Also, the introduction of metal atoms adds all sorts of interesting possibilities, and these are best left to the texts on organometallic compounds.

The title of the book presented difficulties. *Aromatic Compounds* is misleadingly simple or simply misleading. If one takes benzene as *the* aromatic compound then some of the compounds treated are certainly *not* aromatic; they may even be anti-aromatic. The title provides a factual statement of the contents, while the sub-title sums up the approach of the book.

The exact nature of what is included obviously depends upon my own choices and prejudices. I can only hope that it gives the reader the feel of the subject, and develops its methodology adequately. I shall be particularly pleased if it can convey to the reader the delight and interest I have gained and continue to gain from what I find to be an absorbing topic. There are bound to be mistakes (there always are); I can only hope that they are not too numerous. I must thank my colleague and friend Dr Christopher Glidewell for reading much of the text and erasing some errors and suggesting improvements. Finally I must thank members of the staff of John Wiley & Sons for their kind interest and help throughout the preparation of the book.

My main hope is that the reader may enjoy this chemistry.

St Andrews, Fife
May 1989

1

Benzene ([6]Annulene) and Related Compounds

Benzene

Benzene is a unique and really rather strange hydrocarbon. If you had been taught about only aliphatic and alicyclic compounds, and were then suddenly presented with the formula of benzene and asked to predict its properties, you would certainly get them wrong. You would probably suggest that it would readily add bromine, undergo polymerization by the Diels–Alder reaction, and be readily oxidized. From its formula it is obviously an unsaturated compound. Such compounds undergo addition reactions, some with electrophiles, some with nucleophiles. Benzene reacts with electrophiles, but none too readily, and not to give addition products. You would probably be equally unsuccessful in predicting its physical properties.

Let us assume that as a good chemist you accept the experimental evidence and agree that benzene really is stable in air, does not polymerize and undergoes electrophilic substitution reactions. Then let us suppose that in your next examination you are asked to predict the properties of cyclooctatetraene, the next higher homologue of benzene. Not to be caught twice, you will suggest that, as a homologue, it will resemble benzene—it will not add bromine and will be stable to oxidation and polymerization. Wrong again! Cyclooctatetraene might be called a super-alkene; anything that ethylene can do it can do better, and sometimes with some extra complications as well. The chemistry of completely conjugated cyclic polyalkenes is evidently not straightforward. In consequence, such compounds have been the focus of much investigation, both experimental and theoretical.

It is as well to start the consideration of this family of compounds with benzene, since it is the commonest of them and plays such a central role in organic chemistry. Historically, too, it has been known and studied longer than any other cyclic polyalkene.

Annulene Nomenclature

Cyclic polyalkenes which are fully conjugated around the ring have been called *annulenes*. This name is applicable whether or not there is electronic delocalization around the ring. Individual compounds are designated as [*n*]annulenes, where *n* indicates the number of carbon atoms making up the ring.

By this system of nomenclature, benzene and cyclooctatetraene may be called, respectively, [6]annulene and [8]annulene. The annulene names are not commonly used for these two compounds, but they are for annulenes with larger rings. It is simpler to describe a compound as [18]annulene than as cyclooctadecanonaene.

Chemical Properties of Benzene

The most common chemical reactions of benzene are those with electrophiles, e.g. nitration, sulphonation, halogenation, acylation. They will not be described in detail here, since they are common in standard texts of organic chemistry. Suffice it to note that the usual pattern of such reactions is as follows (E^+ = electrophile):

The initial step resembles that of the addition reactions which alkenes undergo, e.g.

$$CH_2{=}CH_2 \xrightarrow{Br_2} CH_2Br{-}\overset{+}{C}H_2 + Br^- \rightarrow CH_2Br{-}CH_2Br$$

The striking difference is that in the case of alkenes the intermediate carbenium ion adds to an anion to give an overall addition reaction, whereas in the case of benzene the intermediate carbenium ion loses a proton to give an overall substitution reaction.

The difference between the behaviour of benzene and that of alkenes cannot be attributed to the delocalization of the positive charge in the intermediate ion derived from benzene as compared to the localized charge in the ion derived from an alkene. A conjugated diene also provides an intermediate ion with a delocalized charge, but the overall reaction of the diene is an addition, not a substitution reaction:

$$
\begin{array}{ccc}
\underset{\text{CH}}{\overset{\text{CH}_2}{\Big\backslash}}\underset{\text{CH}_2}{\overset{\text{CH}}{\Big/}} & \xrightarrow{\ \text{Br}_2\ } & \underset{\underset{\delta^+}{\text{CH}}}{\overset{\text{BrCH}_2}{\Big\backslash}}\underset{\underset{\delta^+}{\text{CH}_2}}{\overset{\text{CH}}{\Big/}}
\end{array}
\longrightarrow
\begin{array}{c}
\text{BrCH}_2\text{CH}{=}\text{CHCH}_2\text{Br} \\
+ \\
\text{BrCH}_2\text{CHBrCH}{=}\text{CH}_2
\end{array}
$$

Nor can the complete cyclic conjugation in benzene provide the explanation, for cyclooctatetraene (C_8H_8), which is the eight-membered ring homologue of benzene, also reacts by addition. (In this case the reaction is rather more complicated (see Chapter 3). This is also true, however, in the case of buta-1,3-diene, mentioned above.)

It appears that there is a special drive for the benzene structure to be reformed. This was emphasized by Armit and Robinson more than sixty years ago in a now-classical paper (*J. Chem. Soc.*, **127**, 1604 (1925)), in which it was described as the 'tendency to retain the type'. Armit and Robinson ascribed this special stability of benzene to its having what are now described as six π-electrons: 'The explanation is obviously that six electrons are able to form a group which resists disruption and may be called the aromatic sextet.' They went on to suggest the inscribed circle formula for benzene, stating 'The circle in the ring

symbolizes the view that six electrons in the benzene molecule produce a stable association which is responsible for the aromatic character of the substance' (i.e. the way in which its properties differed from those of an alkene). Armit and Robinson also commented that 'The unsaturated nature of cyclooctatetraene suggests that a stable group of eight electrons analogous to the aromatic sextet is not formed'.

In 1931 Hückel postulated that 'amongst fully conjugated, planar, monocyclic polyolefines only those possessing $(4n+2)$ π-electrons, where n is an integer, will have special stability' (*Z. Physik*, **70**, 204 (1931); **72**, 310 (1931)). This proposition, known as Hückel's rule, has underlain most subsequent thinking about fully conjugated cyclic molecules.

Structure of Benzene

To return to benzene, the inscribed circle formula implies a D_{6h} symmetry for the molecule. Physical evidence for this symmetry came first from studies of the infrared and Raman spectra. Subsequently electron diffraction studies of benzene vapour also indicated that benzene has a regular planar hexagonal structure with sides of length 1.395 Å, i.e. intermediate between the lengths of single and double carbon–carbon bonds.

Further evidence for the special nature of benzene, as compared to alkanes and alkenes, comes from its ^{1}H-n.m.r. spectrum. It provides a single signal at $\delta 7.27$; the chemical shift indicates that the protons are strongly deshielded, and this has been attributed to a diamagnetic ring current which is induced in the molecules by the applied magnetic field. (For details see one of the many texts available on n.m.r. spectroscopy.) Compounds whose n.m.r. spectra give evidence for diamagnetic ring currents are described as *diatropic*.

Note, however, that, while all the above findings are in accord with benzene having D_{6h} symmetry, they do not completely rule out the possibility that it exists as a molecule with D_{3h} symmetry, i.e. a 'Kekulé structure', which undergoes very rapid interconversion into the other possible valence-isomeric form. Distinction between these two possibilities could be extremely difficult to achieve. Normal spectroscopic, crystallographic or electron diffraction studies would not distinguish between D_{6h} and very rapidly interconverting D_{3h} forms.

The chemistry of benzene is not discussed in detail here because it is covered in all overall textbooks of organic chemistry. Suffice it to say that the benzene structure appears to be one of considerable stability and persistence. The molecular orbital picture of benzene portrays it as having six π-orbitals, of which three are bonding and three are anti-bonding:

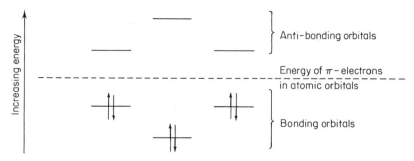

In benzene there are no unoccupied positions in the bonding orbitals, hence entry of a further electron requires expenditure of energy. Furthermore, all the π-electrons are in bonding orbitals, and hence removal of any of them also requires expenditure of energy. Thus benzene, in the ground state, possesses a stable system of six π-electrons, in which either removal of electrons or addition of further electrons would increase the total energy of the system and consequently diminish its stability. This system of six π-electrons in benzene may be likened to the especially stable electron shells of inert gases, and explains its special stability.

Derivatives of Benzene

Whenever one of the hydrogen atoms of benzene is replaced by a substituent group the symmetry of the molecule is lowered. This is reflected in physical properties of the compounds. For example, the remaining hydrogen atoms may no longer provide one singlet signal in their ^{1}H-n.m.r. spectrum but a number of signals with different chemical shifts, depending upon their structural relationships (*o, m* or *p*) to the substituent group(s). Similarly, the structure of the ring may no longer be that of a symmetric hexagon, and bond lengths and bond angles in the ring may not all be identical. For example, in nitrobenzene the bond lengths are as follows:

$$O_2N \quad \overset{\displaystyle 1.43}{\underset{1.37 \quad \diagup \quad \diagdown \quad 1.36}{}}$$

and the internal angle in the ring at C(1) is 125°. However, the overall picture for the great majority of benzene derivatives is that they do resemble benzene closely in their structure. Their ^{1}H-n.m.r. signals generally fall in the appropriate region of chemical shifts, and their bond lengths, if not identical, are all in a range intermediate between the lengths of alkane C—C and alkene C=C bonds.

Chemical properties of benzene derivatives may differ appreciably from those of benzene itself. For example, the unreactivity of benzene towards oxidizing agents, as shown by its failure to decolorize permanganate solutions, does not apply to many benzene derivatives having electron-donating substituents, which may be oxidized by quite gentle oxidizing agents. This is not very surprising—substituted alkanes are very different in their properties from alkanes. One feature runs through all the chemistry of benzene derivatives, however, namely the 'tendency to retain the type'.

Heterocyclic Analogues of Benzene

The simplest type of heterocyclic analogue of benzene is a compound in which one of the carbon atoms of benzene is replaced by a hetero-atom. The archtypical example is pyridine, whose chemical structure has been recognized for well over a hundred years. Pyridine resembles benzene in many ways. For example, their electronic spectra are very similar:

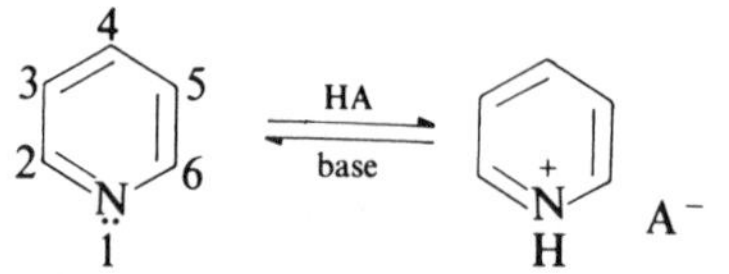

As in benzene, a sextet of electrons is available, each carbon atom and the nitrogen atom providing one apiece. The nitrogen atom also has a lone pair of electrons, not involved in bonding in the ring; this confers a basic character on the molecules, which react with acids to form pyridinium salts. In the formation of these salts the sextet of π-electrons remains. Similarly, N-alkylpyridinium salts are formed when pyridine reacts with alkyl halides.

The presence of the nitrogen atom means that pyridine does not have the D_{6h} symmetry of benzene. In consequence, the ^{1}H-n.m.r. spectrum is not a singlet but consists of three signals associated, respectively, with the 2,6-, 3,5- and 4-positions. The carbon–carbon bond lengths in pyridine closely resemble those in benzene. The carbon–nitrogen bond lengths are intermediate between the lengths of single and double carbon–nitrogen bonds.

Nitrogen is more electronegative than carbon and thus tends to draw the π-electrons towards itself and away from the ring carbon atoms. This is reflected in the fact that pyridine has a dipole moment of 2.2D with the nitrogen as the negative pole.

The electron-withdrawing character of the nitrogen atom, like the presence of similar substituents in a benzene ring, results in deactivation of the ring towards electrophilic substitution. Another and possibly even more important contributing factor to the low reactivity of pyridine towards electrophilic substitution is that the electrophile reacts preferentially with the nitrogen atom to form a pyridinium cation:

Such a cation is naturally resistant to further electrophilic attack.

When electrophilic substitution does take place it may thus involve either reluctant further attack on a pyridinium cation, which will be the predominant species present, or relatively easier attack on the very small amount of uncharged pyridine which is present in equilibrium with the pyridinium cation. Reaction takes place at the 3-position, since this involves the least unfavourable site of attack; attack at the 2- or 4-positions would involve formation of an intermediate with the positive charge partially located on nitrogen, which is energetically less favourable:

Attack at C(2):

Attack at C(3):

The 2,4-positions are also those most deactivated by the electron-withdrawing nitrogen; in a ^{1}H-n.m.r. spectrum (recorded in solution in benzene) the signal for the 3-proton appears at δ 6.78, whereas those for the 2- and 4-protons appear at δ 8.56 and δ 7.10, reflecting the lower electron density at the latter sites. In keeping with this picture, pyridine is not acylated under Friedel–Crafts conditions and is only nitrated under extremely severe conditions.

Replacement of a second carbon atom in the six-membered ring by a nitrogen atom, as in the diazines, results in even more resistance to electrophilic substitution. These compounds have not been nitrated or sulphonated.

Other analogues of benzene having a Group V hetero-atom in place of one of the carbon atoms are known, for example, phosphabenzene and arsabenzene:

Both these compounds are reasonably stable but are sensitive to air. Their photoelectron spectra indicate that the π-bonding is similar to that in benzene and pyridine.

Replacement of a carbon atom by a Group VI element requires the latter to be formally tervalent, i.e. it must be cationic. Such compounds are exemplified by the pyrylium and thiopyrylium salts:

Pyrylium salts are relatively stable but they are also extremely reactive towards nucleophiles, so that they may give a false impression of instability:

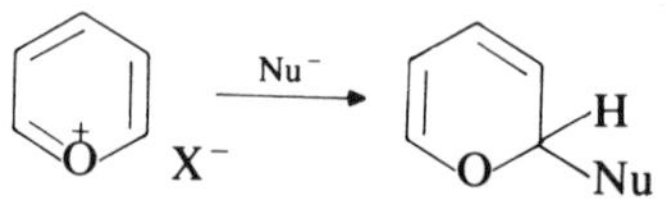

It always needs to be stressed that stability and high reactivity are not incompatible; they represent different features, the inherent energetic stability of a molecule, on the one hand, and a low activation energy for reactions leading to stable products, on the other. Thiopyrylium salts are less reactive towards nucleophiles. For example, thiopyrylium iodide can be recrystallized from hot water, whereas the pyrylium cation reacts rapidly with water even at room temperature. Not surprisingly, simple electrophilic substitution at a pyrylium or thiopyrylium ring has not been described. The electronic spectra of pyrylium cations resemble those of N-alkylpyridinium cations, and their ^{1}H-n.m.r. spectra also show signals in the same region (δ 8–10).

Formulae for Benzene and its Derivatives

Benzene and its derivatives are standardly represented in formulae in two different ways, by the so-called Kekulé formulae (named thus since Kekulé first proposed them) or the inscribed-circle (poached egg) formulae, i.e.

It needs to be emphasized that *both* are only pictograms and that in most cases *neither* provides a completely true picture. The shortcomings of the Kekulé formulae are evident (and were realized by Kekulé); as has been discussed, benzene and its derivatives do not have alternate single and double bonds. However, nor do most benzene derivatives have completely symmetrical cyclic distribution of the π-electrons, as implied by the inscribed-circle formulae. Only in the cases of benzene and its symmetrically substituted derivatives may this be true. Therefore both types of formulae are approximations; even more important, they are symbols or cartoons rather than exact representations, and must be so regarded.

The inscribed-circle formula has particular dangers in the case of polycyclic compounds. For example, if it is used to portray naphthalene

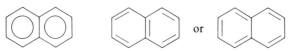

it implies that there are twelve π-electrons in the molecule; in fact there are only ten, so that we *cannot* have two sextets of π-electrons. However, equally, the Kekulé-type formulae can be misleading if they are taken to represent the distribution of electrons within the molecule. So long as it is completely understood that they do not give a precise picture of the electron distribution but are only symbols, Kekulé-type formulae have one great advantage over inscribed-circle formulae for polycyclic compounds in that they do provide an indication of the numbers of π-electrons in the molecules they represent. For this reason, Kekulé formulae are used throughout this book for polycyclic compounds, and frequently for monocyclic compounds as well, and it is assumed that their limitations are recognized.

In Conclusion

We thus have a large family of six-membered ring compounds, heterocyclic as well as carbocyclic, formally represented as having three conjugated double bonds around the ring, which in fact do not have the properties usually associated with molecules having double bonds. Their behaviour has been rationalized in terms of interactions between all the six π-electrons, leading to their cyclic delocalization around the ring. This confers special properties on these compounds, such as ring bond lengths intermediate between the lengths of standard single and double bonds, the induction of diamagnetic ring currents (diatropicity) observed from their ^{1}H-n.m.r. spectra, special stability and electronic spectra.

These special properties have been described as *aromaticity*, and all these compounds have been classified as *aromatic*. In this book these six-membered ring compounds will always be described as *benzenoid*.

The special properties of benzene and its derivatives were recognized more than a century ago. This inevitably led to speculation whether such 'aromatic' properties might also be associated with other fully conjugated cyclic molecules, and this, in turn, inspired a large amount of synthetic work in attempts to prepare these compounds; such work continues today. As mentioned earlier, the first analogue to be prepared, cyclooctatetraene, with four double bonds in an eight-membered ring, was found to be nothing like benzene in its properties. Other compounds have provided both satisfaction and disappointments.

It is with these compounds having conjugated rings which are other than six-membered that this book is largely concerned. As mentioned, a major motivating force in the study of such compounds was to compare their properties with those of benzene. It was inevitable that none would have identical properties, or they would not be different compounds. However, the extent to which they resemble benzene or differ from it in their properties is of great theoretical and practical interest, and this aspect of these compounds will be emphasized throughout the book.

Further Reading

For more information concerning the development of ideas about benzene and its structure see Chapter 1 of the author's 1984 book (for details see Preface).

For an interesting account of the development of the ideas of aromaticity in pre-electronic concepts see J. P. Snyder, in *Non-benzenoid Aromatics*, Vol. I, Academic Press, New York, 1969, pages 1ff.

For a book which is in many ways complementary to the present text, dealing with similar subject matter but often in a different way, see P. J. Garratt, *Aromaticity*, John Wiley, New York, 1986.

For German readers an article by G. Maier '"Aromatisch"—was heisst das eigentlich?', *Chemie in unserer Zeit*, **9**, 131 (1979), is commended.

2
Hückel's Rule

Present-day ideas on cyclic polyalkenes, and on 'aromaticity' are all fundamentally based on Hückel's rule (see page 3). In Hückel's treatment, the π-electrons of the molecules are considered separately from the σ-electrons. The π-electrons of the conjugated system are regarded as common to all the carbon atoms of the system and are considered to occur in molecular orbitals. The molecular orbitals can be of three types: *bonding orbitals*, in which the energy of the electrons is less than their energy in the atomic orbitals; *non-bonding orbitals*, in which the energy of the electrons is equal to their energy in the atomic orbitals; and *anti-bonding orbitals*, in which the energy of the electrons is greater than their energy in the atomic orbitals. If the bonding orbitals are fully occupied the system thereby gains stability, but if either bonding orbitals are unfilled or anti-bonding orbitals are occupied then the stability of the system is greatly reduced. The numbers of each of these kinds of orbitals and their energies for any particular conjugated system depend upon the number of atoms which make up the system and also on the symmetry of the system. The Pauli principle requires that not more than two electrons can be allocated to any one molecular orbital.

Thus this treatment indicates that, as shown diagrammatically in Chapter 1, benzene, in the ground state, contains a stable system of six π-electrons, with no unoccupied positions in the bonding orbitals and no electrons occupying anti-bonding orbitals. Hence either loss or gain of an electron by a benzene molecule requires expenditure of energy.

Hückel's treatment is a simplification of the truth, and has been extensively questioned, but, whatever its theoretical shortcomings, it has had a considerable influence on organic chemistry. It has stimulated most of the work described in later chapters and much more work than is mentioned here. A number of more refined theoretical approaches have been made but will not be discussed. (For details, see B. A. Hess and L. J. Schaad, *Aromaticity*, Wiley–Interscience, New York, in press.)

The representation of the molecular orbitals of benzene on page 4 indicates that there is one orbital of lowest energy and, above this, two other bonding orbitals which are degenerate. Of yet higher energy than these are two degenerate anti-bonding orbitals and finally one further anti-bonding orbital of still higher energy.

The Hückel method indicates that for planar cyclopolyalkenes with rings of other sizes, those having $(4n + 2)$ ring atoms (where n is an integer) have similar symmetric arrangements of orbitals, e.g. for a ten-membered ring

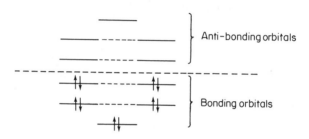

On the other hand, for planar cyclopolyalkene rings having $4n$ ring atoms (n is an integer) the arrangement of the orbitals is different; for example, for cyclobutadiene (C_4H_4):

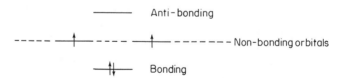

Thus cyclobutadiene should have only two π-electrons in bonding orbitals, with two others in non-bonding orbitals. From Hund's rule these two electrons would be expected to enter separate orbitals, i.e. cyclobutadiene should exist as a diradical.

A useful mnemonic has been devised by Frost and Musulin to illustrate the situation for planar cyclic conjugated polyalkenes. The ring systems are inscribed as regular polygons in circles in such a way that one atom of the ring is at the bottom of the vertical axis. The energies of the π-orbitals associated with the ring displayed correspond to the position of the ring atoms on the circle. Orbitals below the horizontal axis of the ring are bonding orbitals, those above this axis are anti-bonding, while those on the horizontal axis are non-bonding. Thus for benzene and cyclobutadiene we have respectively:

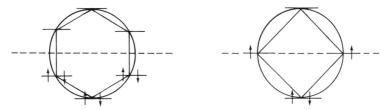

Similarly, for cyclooctatetraene and cyclodecapentaene the Frost and Musulin mnemonics provide the representations:

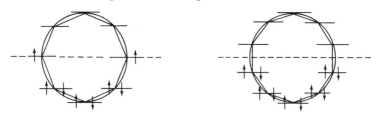

Were this the complete story, cyclobutadiene and cyclooctatetraene would each exist as triplet biradicals. Experimental evidence, as described in later chapters, shows this not to be the case. In fact these molecules distort themselves, thereby lowering the energy of the systems. In so doing they lose their symmetrical geometry, and the two non-bonding orbitals of the symmetric form are no longer degenerate; the two electrons associated with them both occupy the resultant orbital of lower energy. Thus, cyclobutadiene exists as a rectangular molecule with alternate single and double bonds rather than as a square, i.e.

It is interesting to note that other calculations made as long ago as 1937, by Lennard-Jones and Turkevich, also predicted that [4n]annulenes should be made up of bonds of alternate lengths.

Hückel's rule suggests that the annulenes having $(4n + 2)$ π-electrons, where n is an integer, should have special stability. In general terms this rule is still accepted today, although it has been modified and adapted in various ways to take account of more recent experimental and theoretical studies.

In particular, it has been suggested that annulenes having $4n$ π-electrons may not only lack special stability but may indeed suffer destabilization consequent upon delocalization of the π-electrons; they have been described as *anti-aromatic*. The difference between 'aromatic' benzene and 'anti-aromatic' cyclobutadiene has been represented pictorially by energy diagrams such as the following:

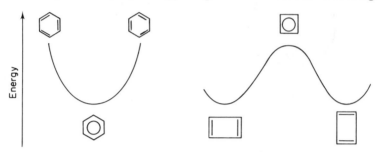

Hence benzene exists in the ground state with a unique geometry, whereas cyclobutadiene is a mixture of valence isomers which require energy for their interconversion.

Note again, however, the point raised in Chapter 1, page 4, that there is no *complete proof* that benzene has D_{6h} symmetry. It is not impossible that it could consist of two very rapidly interconverting isomers, separated by a very small energy barrier, which could give the illusion of a D_{6h} symmetry. The relevant energy diagram would have the form

With such a small energy barrier to interconversion the difference between valence isomerism and mesomerism or resonance becomes trivial. The structural difference between benzene and cyclobutadiene would then be attributable to the size of the energy barrier inhibiting interchange between the two valence isomeric forms in the two cases.

A real problem is associated with the idea of special stability. A guide to the thermodynamic stability of a molecule may be obtained from measurements of its heats of formation or combustion. However, this information may not be easy to obtain, and when it is available, with what does one compare the experimental result? For example, strictly speaking, one should compare values obtained for benzene with those for cyclohexatriene having localized single and double bonds, but, of course, this species is not available.

A popular criterion for aromatic stabilization and anti-aromatic destabilization makes use of Dewar Resonance Energies (DRE) (M. J. S. Dewar, *The Molecular Orbital Theory of Organic Chemistry*, McGraw-Hill, New York, 1969). The energy of an annulene is compared to that of a non-cyclic polyene having the same number of formal double bonds. If the energy of the cyclic molecule is lower than that of the corresponding acyclic polyene, the molecule is regarded as aromatic; if it is higher, it is regarded as anti-aromatic. It must be noted that anti-aromaticity defined in this way does not necessarily indicate an absolute destabilization of the cyclic molecule, only that it is less stable than the corresponding linear polyene.

If the DRE are evaluated for different ring sizes it is found that the values, both positive and negative, decrease rapidly as ring size increases, as shown in the following diagram, and converge to the inherent stabilization which arises from conjugative interaction between adjacent double bonds.

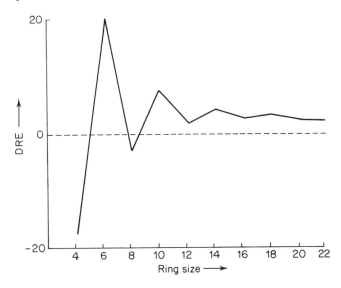

Thus both aromatic stabilization and anti-aromatic destabilization should decrease in larger rings, as should the distinction between the two types of compound.

An experimentally observable distinction between [4n + 2]annulenes and [4n]annulenes involves their ¹H-n.m.r. spectra. It was noted in Chapter 1 that one of the special features of benzene was that the n.m.r. signal for its protons is at $\delta 7.27$, indicating that these protons are strongly deshielded. Annulenes having larger rings may have some of their protons within the ring in the preferred shape of the molecule. This is shown by [18]annulene:

As with benzene, the protons outside the ring are strongly deshielded, providing a signal at $\delta 9.28$. The protons within the ring provide a signal at $\delta - 2.99$, i.e. they are strongly *shielded* by the diamagnetic ring current which is induced in the molecules. [18]Annulene is a [4n + 2]annulene (n = 4); other [4n + 2]annulenes provide similar spectra, with deshielded outer protons and shielded inner ones.

For [4n]annulenes the positions of the signals for the inner and outer protons are reversed. Thus [16]annulene shows signals at $\delta 5.40$ and $\delta 10.43$, the former arising from the outer protons and the latter from the inner ones. This pattern shown by [4n]annulenes is associated with a paramagnetic ring current, which can be rationalized in terms of an unusually small HOMO–LUMO (highest occupied molecular orbital, lowest unoccupied molecular orbital) splitting, which in turn is associated with the Jahn–Teller effect which replaces two non-bonding orbitals by two orbitals, one weakly bonding (HOMO) and one weakly anti-bonding (LUMO). As compounds in which diamagnetic ring currents are induced are called *diatropic*, so those in which paramagnetic ring currents are induced are described as *paratropic*. Cyclic polyalkenes whose ¹H-n.m.r. spectra give no indication of any ring current are described as *atropic*. The extent of the *diatropicity* or *paratropicity* (i.e. the difference in chemical shift between the signals for the inner and outer protons) of annulenes has been used as a guide to the extent of delocalization in the ring.

In addition to stabilization energies and ^{1}H-n.m.r. spectra, a difference between $[4n + 2]$- and $[4n]$annulenes might be expected in the degree of alternation of bond lengths around the ring. It was noted in Chapter 1 that all the ring bonds of benzene are of the same length, this being associated with the delocalization of π-electrons around the ring.* It seems reasonable that $[4n + 2]$annulenes, for which delocalization of π-electrons provides a stabilizing influence, might be expected to have ring bonds of (approximately) similar length, whereas $[4n]$-annulenes, which may be destabilized by delocalization of π-electrons, would have alternate longer and shorter bonds corresponding to localized single and double bonds. In a general way this is the case, but the cautionary note introduced in Chapter 1 (see page 5) needs to be sounded even more strongly when applied to the generality of ring sizes, for various other factors (e.g. steric) must also contribute to the overall structure.

Cyclic Ions

Hückel's rule may be applied to cyclic ions of general formula C_nH_n as well as to neutral molecules.

In 1901 Thiele described the ready deprotonation of cyclopentadiene to form a cyclopentadienide anion of unusual stability:

$$C_5H_6 \underset{\text{H}^+}{\overset{-\text{H}^+}{\rightleftharpoons}} [C_5H_5]^-$$

This anion is symmetric as shown by its n.m.r. spectra, and must have six electrons not involved in the σ-framework of the molecule, two from each of the double bonds of cyclopentadiene and two from the C—H bond which is cleaved to form the anion.

Hückel's rule predicts that this anion should be stabilized by formation of a delocalized sextet of electrons; a similar explanation had earlier been provided by Robinson and Armit in terms of their concept of the 'aromatic sextet'. This ion is depicted as

*In recent years there have been suggestions that this approach puts the cart in front of the horse, and that the delocalization of π-electrons follows from the symmetric nature of the σ-framework (see various papers by S. S. Shaik, P. C. Hiberty *et al.* and K. P. C. Vollhardt (1985 onwards); their views are not accepted by some other chemists. It is always possible that it is not a horse/cart situation but a chicken/egg one.

The circle symbolizes the sextet of π-electrons and the negative sign within the circle indicates that the negative charge, like the π-electrons, is shared over the whole ring.

The relevant Frost–Musulin diagram also suggests the stability of this ion, with all the electrons being in bonding orbitals:

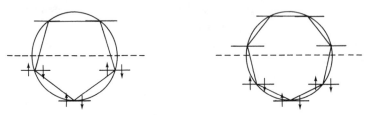

The neighbouring Frost–Musulin diagram suggests that in the case of the corresponding seven-membered ring (C_7H_7), formed by removal of hydrogen from cycloheptatriene (C_7H_8), the stable species should also be one having six π-electrons, i.e. formed by removal of a hydride anion to provide a cation $[C_7H_7]^+$. The anion $[C_7H_7]^-$ would be expected to have two electrons in anti-bonding orbitals.

Hückel predicted the formation of the stable $[C_7H_7]^+$ ion in 1931. In fact it had already been prepared in 1891, but its structure was not realized, and in 1954 the work was repeated and Hückel's prediction was confirmed.

Both the cyclopentadienide anion, $[C_5H_5]^-$, and the cycloheptatrienium cation, $[C_7H_7]^+$, more commonly known as the *tropylium* ion, give ^{1}H-n.m.r. spectra indicative of their delocalized character and symmetry (each appears as one singlet), and the presence of a diamagnetic ring current. The chemical shifts, at $\delta 5.5$ for $[C_5H_5]^-$ and at $\delta 9.2$ for $[C_7H_7]^+$, differ from that of benzene ($\delta 7.27$) because of the negative and positive charges which these ions respectively carry.

These ions thus resemble benzene structurally and spectroscopically. Not surprisingly, in view of the charges they carry, their chemical properties show less resemblance.

A characteristic property of benzene is its ability to react with electrophiles. It is not surprising that the tropylium ion is not attacked by such reagents since it bears a positive charge, and is electron-deficient with six π-electrons spread over seven ring atoms.

Another characteristic property of benzene is its unreactivity towards our atmosphere; it can be kept in contact with air containing oxygen and water without change. In contrast, the cyclopentadienide ion is extremely sensitive to air and reacts violently with oxygen or water. Cyclopentadienide salts rapidly resinify in air and may inflame; even the less reactive thallium salt is oxidized in air so vigorously that it chars paper on which it is placed. Yet these salts are thermodynamically stable; sodium cyclopentadienide remains unchanged when heated to 300 °C in nitrogen for a long period.

Thus it is well to emphasize once again that chemical reactivity and instability are not synonymous. Confusion between instability and reactivity is a common pitfall in chemistry; very stable molecules can be very reactive in the appropriate conditions.

Benzene, the tropylium ion and the cyclopentadienide ion all share the common features of thermodynamic stability and the tendency to retain the type. Even the highly reactive cyclopentadienide ion may retain the type under appropriate conditions; not surprisingly, it is highly reactive towards electrophiles.

In succeeding chapters the chemistry of neutral cyclopolyalkenes and of related ionic species will be discussed. Inevitably, we shall frequently be looking over our shoulders at benzene and its derivatives to make comparisons and contrasts. However, this is not the only objective of this text. Rather, the various groups of compounds will be considered for their own sakes, and different aspects of interest of their chemistry will be described. Their modes of preparation will be outlined early in each chapter. Most of these compounds cannot be bought off the shelf and the ways in which they are obtained are of interest. This will be followed by accounts of their structure, spectra and chemical behaviour. Most of this information is not obtainable or is only mentioned very superficially in standard textbooks on organic chemistry, yet it has much fundamental interest and teaches us a lot about basic organic chemistry.

Further Reading

There are various textbooks available on molecular orbital theory. For information about Dewar Resonance Energies see M. J. S. Dewar, *The Molecular Orbital Theory of Organic Chemistry*, McGraw-Hill, New York, 1969.

For more detail about the Hückel method see P. J. Garratt, *Aromaticity*, John Wiley, New York, 1986, Chapter 1.

3
Cyclooctatetraene

We will consider cyclooctatetraene next because it was the first annulene after benzene to be prepared and described. Its initial preparation, by Willstätter *et al.* (1911, 1913), started from the alkaloid ψ-pelletierine, which was obtained from pomegranate bark. ψ-Pelletierine had the advantage of having an eight-membered ring already present in its molecules:

ψ-Pelletierine

A series of steps reduced the carbonyl group to a hydroxyl group which was then eliminated to provide the first double bond. Successive exhaustive methylation procedures on the nitrogen atom both removed this nitrogen bridge and introduced two further double bonds into the eight-membered ring. Successive treatment of the resultant cyclooctatriene with bromine and dimethylamine, followed by further exhaustive methylation, provided cyclooctatetraene as a yellow liquid.

It was not like benzene in its properties. It polymerized when heated or just on being kept at room temperature. The lack of 'aromatic' properties surprised most organic chemists at the time, conditioned as they were by the properties of a cyclic system of formal alternate single and double bonds, as in benzene. To rationalize their surprise some chemists questioned the work and disputed the structure of the product; most objections were based on the apparent close similarity between the properties of the product and those of styrene.

In 1945 it was found that during the Second World War a simple synthesis of cyclooctatetraene had been devised in Germany, and it had in fact been produced in substantial amounts. This was done by tetramerization of acetylene in furan as solvent with a nickel salt present as catalyst:

$$4C_2H_2 \quad \xrightarrow[\text{NiCl}_2 \text{ or Ni(CN)}_2]{\text{Heat, pressure}} \quad$$

Shortly afterwards a further multi-step synthesis was carried out, starting from chloroprene (2-chloro-buta-1,3-diene), which was first dimerized to a dichlorocyclooctadiene.

All three synthetic routes provided the same product and the authenticity of the original sample was vindicated. Indeed, the original synthesis was repeated and found to be correct in all particulars.

Structure

Since cyclooctatetraene has a cyclic array of eight π-electrons there is no reason to expect any special stabilization arising from annular conjugation of these π-electrons. Indeed, any such conjugation would necessitate coplanarity of the eight-membered ring, which would result in the introduction of angle strain into the molecules. If, however, there is no energetic gain to be had from the molecules taking up planar structures, there is every reason for them to take up rather a tub shape, which largely eliminates angle strain and reduces non-bonded repulsive interactions between the hydrogen atoms:

Electron diffraction measurements show that there are alternate single and double bonds of length 1.462 and 1.334 Å. The infrared spectrum is typical of a non-conjugated alkene. The hydrogen atoms are all equivalent and provide a single signal in the ^{1}H-n.m.r. spectrum, in the alkene range, at $\delta 5.68$.

Cyclooctatetraene molecules take part, however, in two dynamic processes, one involving inversion of the ring

and the other valence isomerism, i.e. exchange of the positions of the single and double bonds which may be represented as follows:

Yet another dynamic process involves valence isomerization between cyclooctatetraene and bicyclo[4.2.0]octatriene:

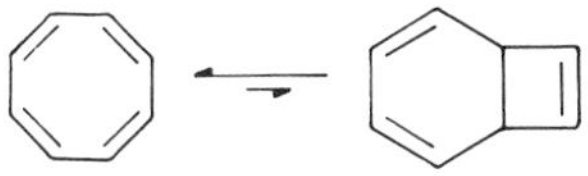

This is an example of a thermal 6π electrocyclic equilibration. The bicyclic form is present in only very small amounts ($\sim 0.01\%$ at $100\,^{\circ}C$), but, as will be seen later in this chapter, it plays a role in the chemistry of cyclooctatetraene. It is interesting to note that steric factors may affect these equilibria. An extreme example is that of the following di-t-butyl derivative, which exists entirely in the bicyclic form:

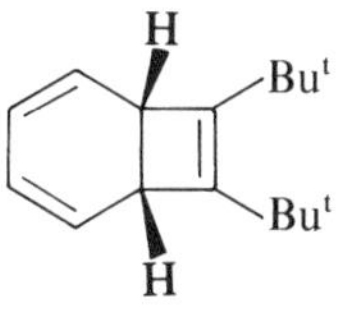

This compound does not isomerize to di-t-butylcyclooctatetraene, presumably because of the increased steric strain which would result.

Chemical Reactions

Cyclooctatetraene can be reduced catalytically. Reduction of the first three double bonds proceeds much more rapidly than that of the fourth and last double bond, so that it is possible to obtain either cyclooctene or cyclooctane as product by stopping the reduction after the appropriate amount of hydrogen has been taken up:

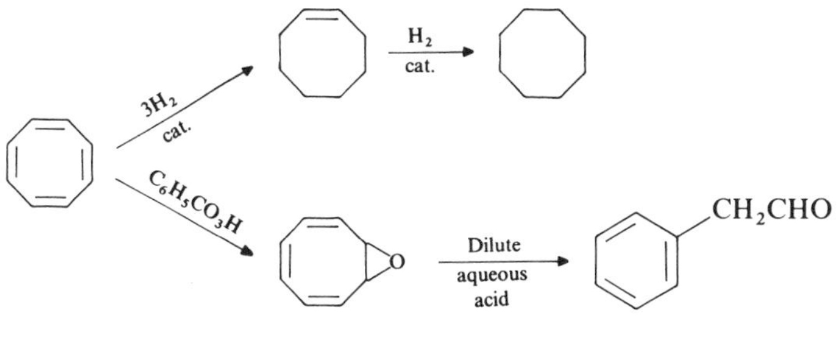

Oxidation with per-acids provides a monoperoxide. This epoxide rearranges in the presence of aqueous acid to give phenylacetaldehyde. Other oxidative procedures lead to benzene derivatives as the isolated products. Thus phenylacetaldehyde is obtained by the action of mercury(II) salts, while sodium hypochlorite or chromium trioxide provide, respectively, terephthalaldehyde and terephthalic acid:

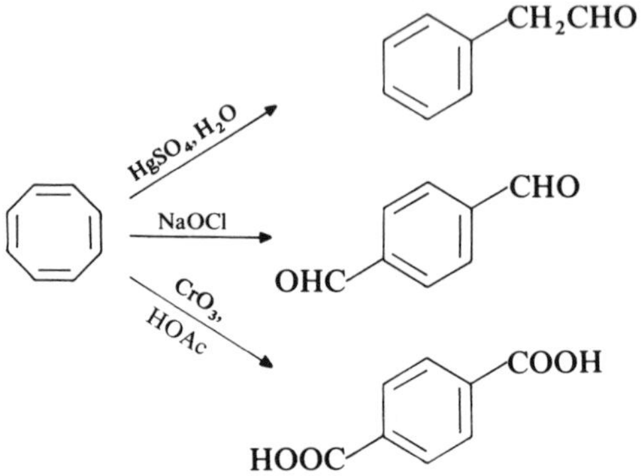

Reaction of cyclooctatetraene with either chlorine or bromine leads to addition products which are bicyclo[4.2.0]octadiene derivatives:

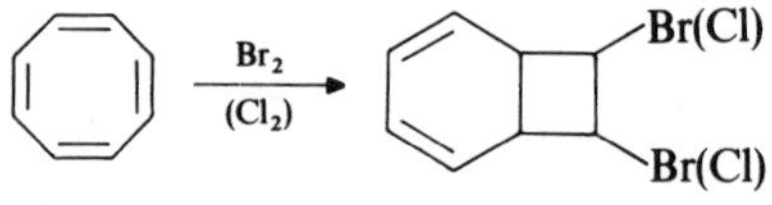

Rearrangement of the eight-membered ring to the bicyclic structure appears to be the last step in this reaction, involving valence isomerization of an initially formed dihalogenocyclooctatriene.

Bicycloocta[4.2.0]diene derivatives also result when cyclooctatetraene is treated with reactive dienophiles, e.g.

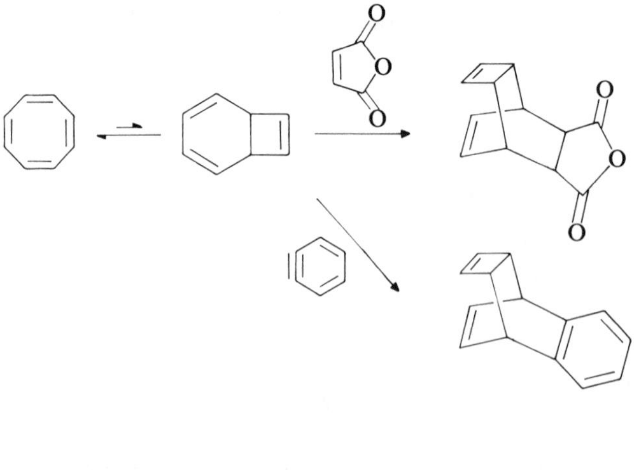

In this case, however, the monocyclic–bicyclic rearrangement precedes the addition reaction, and the cyclo-addition reaction involves the valence isomer of cyclooctatetraene rather than cyclooctatetraene itself.

The converse of this monocyclic–bicyclic isomerism can be utilized to obtain cyclooctatetraene derivatives from initially formed bicycloocta[4.2.0]diene derivatives, e.g.

When cyclooctatetraene became readily available it was observed that, when treated with alkali metals, it gave alkali metal derivatives which, with water, gave a mixture of cyclooctatrienes. These metal derivatives are discussed in the following section.

Cyclooctatetraenide Salts

In 1956 it was suggested, on the basis of magnetic susceptibility measurements, that the dilithium derivative might be a salt of a cyclo-octatetraenide dianion, which would have a system of ten π-electrons, i.e. in terms of Hückel's rule it might be a stabilized system.

In 1960 a dipotassium salt was isolated as very pale yellow crystals:

This salt was thermodynamically stable in crystalline form or in solution and could be kept in an inert atmosphere, but it was also extremely reactive towards oxygen or moisture, so much so that it exploded on exposure to air.

Its ^{1}H-n.m.r. spectrum consists of a single sharp peak at $\delta 5.7$ and its infrared spectrum is very simple. Both spectra are in accord with the anion being planar with a delocalized decet of π-electrons, and with D_{8h} symmetry. A dianion might be expected to provide a ^{1}H-n.m.r. signal at a lower δ-value; the observed chemical shift indicates that the effect of charge is opposed by the effect of a diamagnetic ring current induced in the ring. The ^{7}Li-n.m.r. spectrum of the lithium salt suggests that in solution in tetrahydrofuran this salt exists as ion-contact pairs, with the lithium ions co-ordinated with the π-electrons of the ring.

An X-ray structure determination has been made of a diglyme complex of dipotassium 1, 3, 5, 7-tetramethylcyclooctatetraenide. In the crystal the anion is flat with fourfold symmetry. The carbon–carbon bonds in the ring average 1.407 Å in length. More recently, an X-ray structure of the tetrahydrofuran complex of dipotassium cyclooctatetraenide ($C_8H_8K_2 \cdot 3THF$) has also been reported. In this case also the ring is planar and regular with bond lengths as in the tetramethyl derivative.

In the cyclooctatetraenide dianion the increased energy required to achieve planarity is outweighed by the energetic gain in the delocalization of the ten π-electrons, especially associated with the delocalization thereby of the double negative charge.

The cyclooctatetraenide dianion can participate in reactions either as a reducing agent or as a nucleophile. Examples of such reactions are:

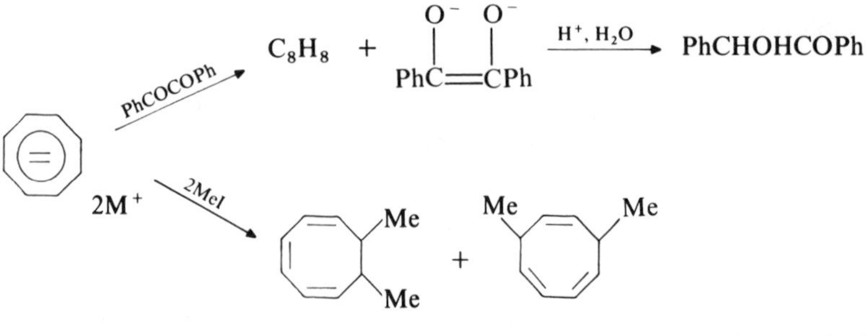

Many of its reactions with nucleophiles are more complex and give rise to a mixture of products. The reaction with acetyl chloride may be taken as an example:

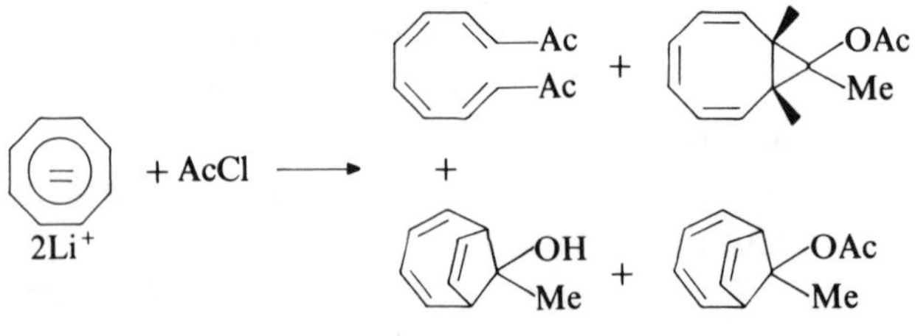

With geminal dichlorides bicyclo[6.1.0]nonatrienes are formed, e.g.

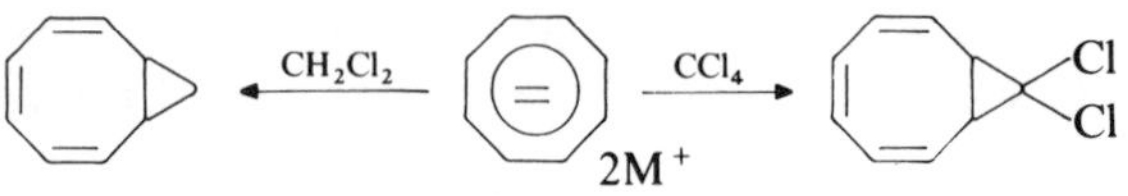

The dianion forms complexes with a number of metals. An example is uranocene, which forms green crystals. These are thermally stable, but inflame in air. X-ray studies show that it has a sandwich structure, with the uranium atom equidistant from all the carbon atoms, and with carbon–carbon bonds all of equal length (1.395 Å):

Dehydrocyclooctatetraenes

Removal of pairs of neighbouring hydrogen atoms from cycloocta-tetraene would provide compounds in which one or more of the alkene bonds were replaced by alkyne bonds, e.g.

These compounds may be called dehydrocyclooctatetraenes. Replacement of the double bonds by triple bonds forces part or all of the ring to become coplanar. Electronic interaction between the multiple bonds should be similar to that in planar cyclooctatetraene; the additional π-bond of the triple bond(s) is orthogonal to the other π-bond. However, much strain is introduced into the structure, both because of the enforced planarity of the ring and because the bonds adjacent to the triple bond are constrained by the ring to be non-colinear. Cyclooctatrienyne has been prepared in solution by treatment of bromocyclooctatetraene with a base. It was not isolated but was trapped by a number of reagents.

Heterocyclic Analogues of Cyclooctatetraene

A number of azacyclooctatetraenes are known, e.g.

Whereas the 1,2-diaza-derivative exists only as the one valence bond isomer, the 1,4-diaza-derivative shown exists as a mixture of the two valence-bond isomers. With potassium the latter compound provides a diatropic delocalized dianion.

1,4-Diazacycloocta-2,5,7-trienes might also have diatropic delocalized ten π-electron systems:

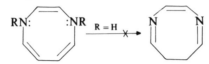

The parent compound, R = H, very sensitive in solution to air. It is stable towards both acid and base, and does not revert to the isomeric structure having two methylene groups, despite the fact that change of an enamine structure into an imine one is usually energetically favoured. From this observation, and from its ^{1}H- and ^{13}C-n.m.r. and electronic spectra, it was concluded that the ring was planar with delocalization of the π-electrons. X-ray studies confirm the planarity of the rings in the crystalline state, with some equalization of the bond lengths, although less than might be expected.

Substituted derivatives in which R represents an electron-donating group (e.g. methyl) are also planar and diatropic, but when R is an electron-withdrawing group (e.g. SO_2R, COOMe, COPh) the ring is twisted, the bonds are alternately double and single, and the compound is atropic. This latter case presumably reflects competitive interactions of the nitrogen lone pairs of electrons with the N-substituent rather than with the π-electrons of the alkene bonds. The fine balance between delocalization or localization of the π-electrons, depending on the nature of the substituents, is worth noting; the energetic difference between delocalized and localized structures is not infrequently small. The dipotassium salt (R = K) has also been obtained in solution and is diatropic.

Related 1-oxa-4-azacycloocta-2,5,7-trienes have been prepared and show similar behaviour, in that if a hydrogen atom or electron-donating group is attached to the nitrogen atom the ring is planar and diatropic, but if an electron-withdrawing group is attached to the nitrogen atom it is non-planar and atropic. Not surprisingly, for both the oxa-aza and diaza compounds, electron-withdrawing substituents on the nitrogen atom(s) make them less sensitive to air; the other examples are very sensitive to air.

The properties of the 1,4-dioxa analogue indicate that it is entirely olefinic in character. This compound was prepared by isomerization of diepoxycyclohexene:

The equilibrium strongly favours the dioxacyclooctatetraene. In contrast, the analogous dithioepoxycyclohexene does not isomerize to give dithiacyclooctatetraene.

Monoaza- and monooxacyclooctatetraenide salts have been prepared, using strong non-ionic bases:

$X = Bu^tN, TosylN, O$

N.m.r. spectra show that the anions are planar and diatropic. The oxa-dianion is much less stable than the aza-dianions.

Further Reading

For more details of the chemistry of cyclooctatetraene and its derivatives the reader is referred to G. I. Fray and R. G. Saxton, *Chemistry of Cyclooctatetraene and its Derivatives*, Cambridge University Press, Cambridge, 1978. An article on cyclo-octatetraenide salts, by M. Sauerbier and H. Kolshorn, appears in *Methoden der Organischen Chemie (Houben–Weyl)*, 4th edition, Vol. 5/2c, pp. 86ff., Thieme, Stuttgart, 1985.

See also Chapter 6 of the author's 1984 book (for details see Preface).

4

Cyclobutadiene

One of the longest-running sagas of organic chemistry was that of the attempted preparation of cyclobutadiene, a molecule with a deceptively simple structure:

An attempt to prepare cyclobutadiene-1,2-dicarboxylic acid, by the action of alkali on 1,2-dibromocyclobutane-1,2-dicarboxylic acid, was published as long ago as 1894, but only cyclobutene derivatives were obtained. About ten years later unsuccessful attempts were made to prepare cyclobutadiene itself, starting from 1,2-dibromocyclobutane. Another sixty years elapsed before cyclobutadiene was eventually obtained, being trapped *in situ* by reaction with dienophiles.

Contemporary theoretical insight makes this inaccessibility appear very reasonable. The early workers did not have our advantage of hindsight and, as with cyclooctatetraene, the example of benzene as a 'cyclic polyene', and the apparent simplicity of the target molecule, made the synthetic attempts appear very reasonable.

It is now realized that at least two factors contribute to the difficulty of isolating cyclobutadiene; both lead to its having very high chemical reactivity. As discussed in Chapter 2, annulenes having $4n$ π-electrons may suffer destabilization due to delocalization of the π-electrons, and, as benzene is *the* 'aromatic' compound, so cyclobutadiene may be considered to be *the* 'anti-aromatic' compound. Also, as discussed in Chapter 2, if the molecules escape from delocalization of the π-electrons by taking up rectangular shapes in which there is diminished interaction between the π-electrons of the two resultant isolated double bonds, cyclobutadiene still represents a highly reactive species which undergoes cyclo-addition reactions with the greatest of ease, with dimerization ensuing if no other reactant is available:

Cyclobutadiene also suffers from considerable angle strain; this is partly relieved in addition reactions, since the presence of double bonds introduces markedly more strain into a four-membered ring. Angle strain was once thought to be the main contributing factor to the instability or reactivity of cyclobutadiene. Later it appeared that this was not the overriding problem, since other compounds having four-membered rings made up of four trigonal carbon atoms, dimethylenecyclobutenes and tetramethylenecyclobutanes, e.g.

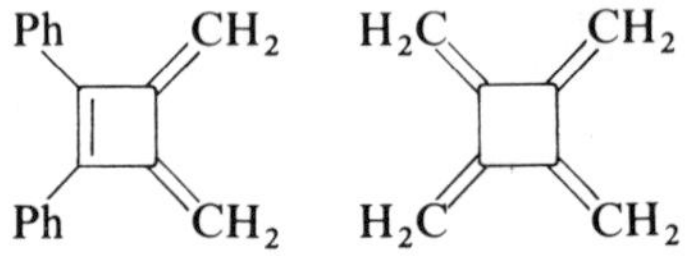

were isolable, although they did polymerize rapidly at room temperature.

More Concerning the Preparation of Cyclobutadienes

An attempted dechlorination of 3,4-dichloro-1,2,3,4-tetramethylcyclobutene with lithium amalgam, carried out in 1957, led to the formation of a tricyclic compound, which could very plausibly arise by dimerization of the tetramethylcyclobutadiene which had been the target of the reaction:

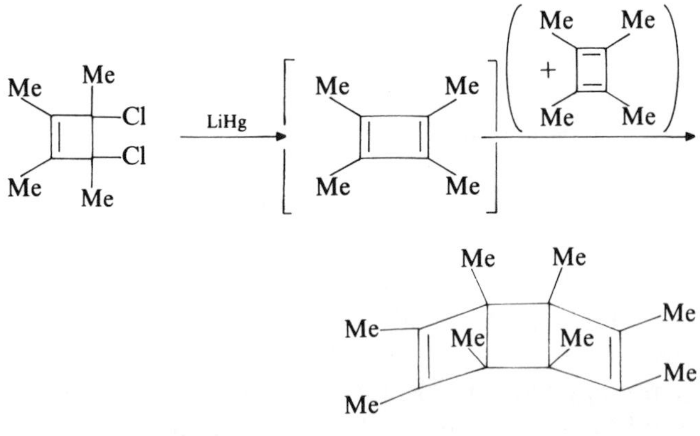

If the octamethyltricyclooctadiene is indeed formed by such a cycloaddition reaction, then the intermediate tetramethylcyclobutadiene should also be trappable by some other reactive dienophile. This was achieved some years later by treating the dichloro compound with zinc in the presence of either dimethylacetylene or dimethyl acetylenedicarboxylate. The isolated products were, respectively, hexamethylbenzene and dimethyl tetramethylphthalate, which could arise from ring-opening of the initial adducts to provide the benzene derivatives; such reactions are well established:

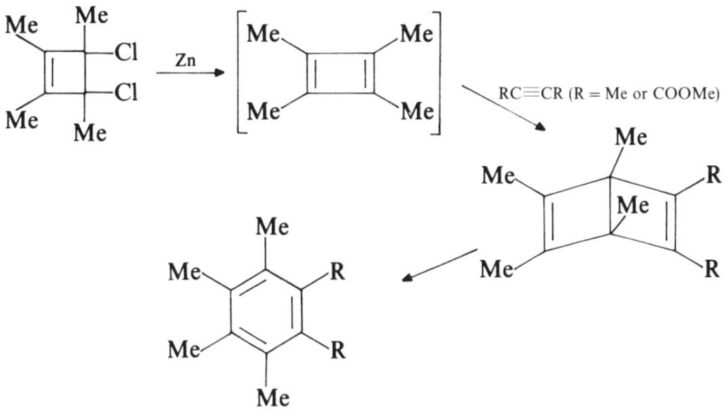

If was suggested by Longuet-Higgins and Orgel in 1956 that cyclobutadiene should form stable complexes with transition metals. Soon afterwards reddish-violet crystals of a nickel complex of tetra-methylcyclobutadiene, $[C_4Me_4NiCl_2]_2$, were prepared. They were soluble in water and in chloroform and the ^{1}H-n.m.r. spectrum showed that all the hydrogen atoms were equivalent. When heated under reduced pressure this complex decomposed to give a dimer of tetramethylcyclobutadiene.

Oxidation of a stable iron tricarbonyl complex of unsubstituted cyclobutadiene led to the formation of cyclobutadiene itself in 1965. It was trapped by having a reactive dienophile present, with which it underwent a cyclo-addition reaction; e.g.

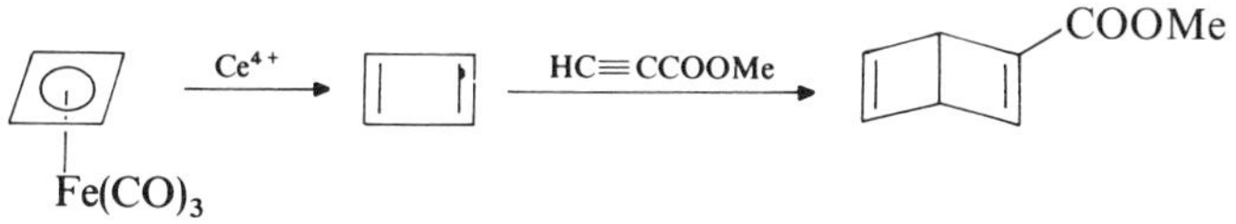

The product readily undergoes rearrangement to give methyl benzoate. If no such trapping agent is present, the cyclobutadiene dimerizes to *syn* and *anti* tricyclooctadiene. This is still the best method available for the preparation of cyclobutadiene. Substituted cyclo-butadienes have been made in a similar way.

An alternative method of preparation of cyclobutadiene involves irradiation of a matrix containing a suitable precursor molecule which can lose stable leaving groups, leaving cyclobutadiene. Examples of this are:

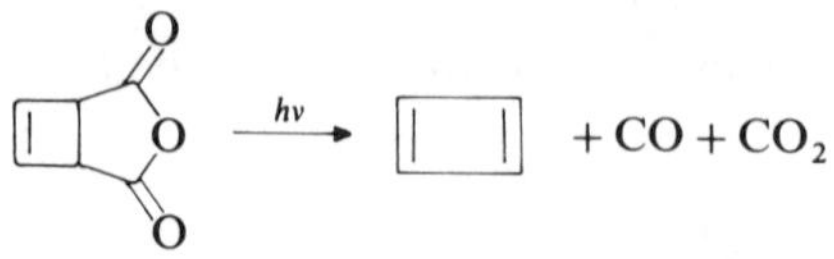

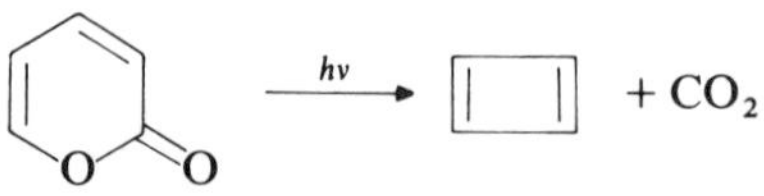

The first of these is the most convenient way of preparing cyclobutadiene in a matrix.

In this method, the precursor is first dispersed through an inert matrix (for example, argon) at low temperature and the matrix is then irradiated. The individual molecules of cyclobutadiene which are generated are themselves dispersed through the matrix, and are thus prevented from reacting with one another. Spectroscopic investigations can be made on the matrix. When the matrix is warmed and melts, the molecules of cyclobutadiene immediately react together to form the dimer. It is not possible to handle cyclobutadiene in the gas phase and collect it on a cold finger because of its extremely high reactivity. Substituted cyclobutadienes have also been prepared in matrices, by irradiation of, for example, appropriately substituted derivatives of α-pyrone and cyclopentadienone.

Stability of Cyclobutadienes

Cyclobutadiene and its derivatives dimerize readily in a $(4 + 2)\pi$ cyclo-addition reaction:

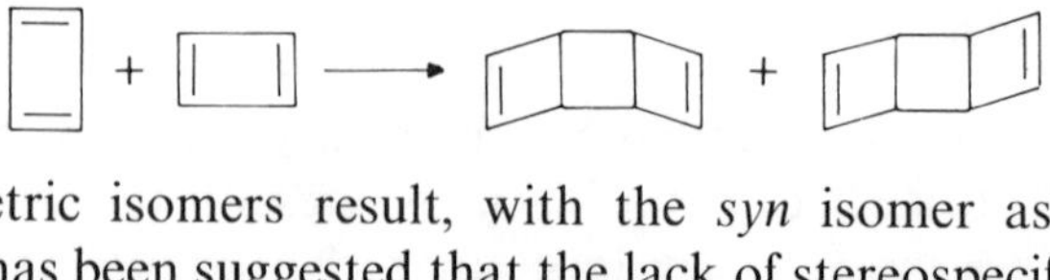

Two geometric isomers result, with the *syn* isomer as the major product. It has been suggested that the lack of stereospecificity, which results in the formation of both products, is a consequence of the very high reactivity of cyclobutadiene both as a diene and as a dienophile.

Increase in the number and size of alkyl substituents on the cyclobutadiene ring decreases its reactivity in cyclo-addition reactions. Thus tri-t-butylcyclobutadiene can survive for some time at room temperature and only changes into the dimer over a period of about a day. This compound is, however, very sensitive to oxidation; in air its rate of autoxidation is faster than its rate of dimerization. Tetra-t-butylcyclobutadiene also survives for some time at room temperature in an inert atmosphere, but is very readily oxidized in air. Sensitivity to oxidation is a general characteristic of cyclobutadienes.

Cyclobutadiene is decomposed photolytically into acetylene but 1,3-dimethylcyclobutadiene is photochemically stable. Irradiation of tetra-t-butylcyclobutadiene results in its isomerization to give tetra-t-butyltetrahedrane:

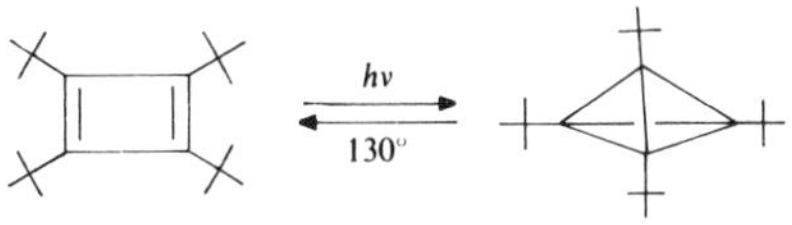

This tetrahedrane is extremely stable in air and is not sensitive to oxygen. It is reconverted into tetra-t-butylcyclobutadiene when it is heated.

Thermally stable cyclobutadiene derivatives have been obtained by having both electron-withdrawing and electron-donating substituents attached to the ring, as in

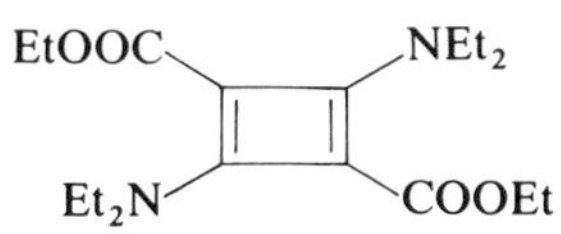

This compound was isolated as a yellow crystalline material. It owes its stability to the 'push–pull' interaction which can take place between the substituent groups, i.e.

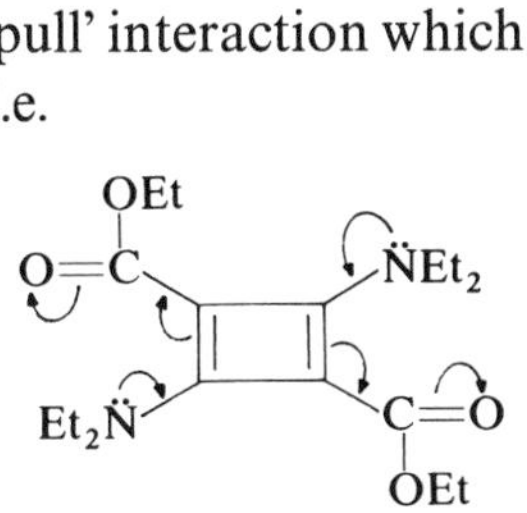

The presence of both donor and acceptor substituent groups in polyalkenes generally increases the LUMO–HOMO gap in $4n$-π-electron systems, and thus increases the possibility of obtaining stable $4n$-systems.

An X-ray study shows that the bonds linking the substituent groups to the ring are shortened, which is consistent with push–pull interactions. Electronic and photo-electron spectroscopy indicate that in the ground state the molecule is most economically described in terms of two push–pull systems which have little interaction with one another, resulting in a rectangular-shaped ring. In contrast, X-ray studies and n.m.r. spectroscopy suggest that all four sides of the ring are of equal length. The likeliest resolution of this discrepancy is that the molecules undergo rapid valence isomerization between two degenerate forms:

EtOOC⎯⎯NEt$_2$ ⇌   EtOOC⎯⎯NEt$_2$
Et$_2$N⎯⎯COOEt   Et$_2$N⎯⎯COOEt

If the energy barrier to interconversion of the valence isomers is low and rearrangement is consequently fast on the X-ray and n.m.r. time scales the square model and the rapidly interconverting rectangular model would give delusively identical electron-density maps. These 'push–pull' cyclobutadienes react readily with both electrophiles and nucleophiles, the former reacting at a ring atom carrying an ester group and the latter at an atom bearing an amino substituent. This reactivity is a consequence of the electronic interaction from the substituent groups which allow the cyclobutadiene to behave respectively as an enamine or as an α,β-unsaturated carbonyl compound. These 'push–pull' cyclobutadienes are sensitive towards water and other protic solvents.

Structure of Cyclobutadiene

Early theoretical treatments of cyclobutadiene suggested that it would have a square planar geometry with two π-electrons in bonding orbitals and two singly filled non-bonding orbitals. The lowest energy state would therefore be a triplet biradical.

Most modern theory predicts a rectangular singlet ground state which is rapidly equilibrating between two possible valence isomeric forms:

Interconversion of these forms proceeds via a square singlet intermediate; calculations indicate that a square triplet form is of even higher energy. The relative energies can be summed up by the following energy diagram:

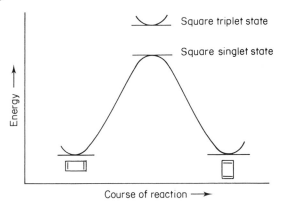

The height of the energy barrier is $\sim 10\,\text{kcal mol}^{-1}$.

It is possible that when cyclobutadiene is generated photochemically in a matrix at low temperature it may be in an excited triplet state which, at these low temperatures, is metastable and converts only slowly into its singlet state.

X-ray analyses of sterically hindered, and hence sterically crowded, cyclobutadienes indicate that in each case the molecules are rectangular; ring bond lengths (Å) are as shown:

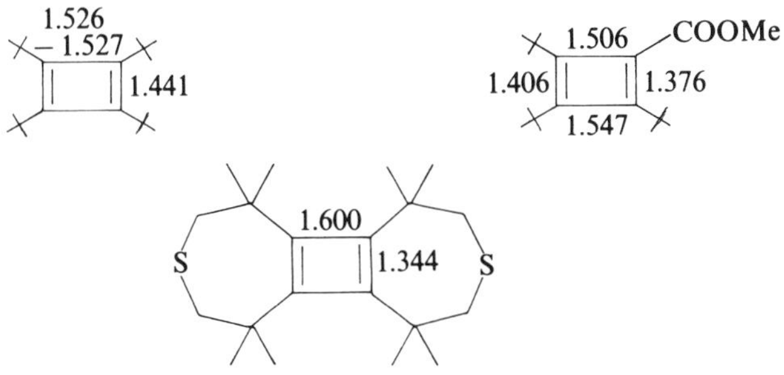

The smaller differences in bond lengths in the t-butylcyclobutadienes probably arise from steric effects due to the bulkiness of the t-butyl groups. There is evidence that the tetra-t-butyl derivative may be in rapid equilibrium with its valence isomer at room temperature; the bond lengths quoted were obtained at 123 K.

N.m.r. studies of other t-butyl substituted cyclobutadienes are also in accord with structures consisting of rapidly equilibrating valence isomers. Variable-temperature studies of the tri-t-butyl derivative indicate that the valence isomers cannot be frozen out down to $-190\,^\circ$C, and that the activation energy for isomerization must be less than $2.5\,\text{kcal mol}^{-1}$. Photo-electron spectra of tri- and tetra-t-butylcyclobutadienes are also in accord with this picture. ^{13}C-n.m.r. spectra of ^{13}C-labelled cyclobutadiene at $\sim 25\,$K show unequivocally that the spectrum is due to rapidly interchanging valence isomers. Initially, the infrared spectra of cyclobutadienes were interpreted in terms of a square structure, but later analyses are also in agreement with the rectangular form.

Infrared spectra of samples of cyclobutadiene generated in an inert matrix by reactions which also lead to the formation of other gaseous products indicate that the confinement together within the matrix of cyclobutadiene and other product(s) may result in some deformation of the geometry of the cyclobutadiene. Electronic spectra of samples of tetramethylcyclobutadiene prepared in a matrix from various precursors differ from one another; this also indicates intermolecular interactions within the matrix.

The earliest evidence for the singlet nature of cyclobutadiene came from trapping experiments. Either dimethyl maleate or dimethyl fumarate added stereospecifically to the diene:

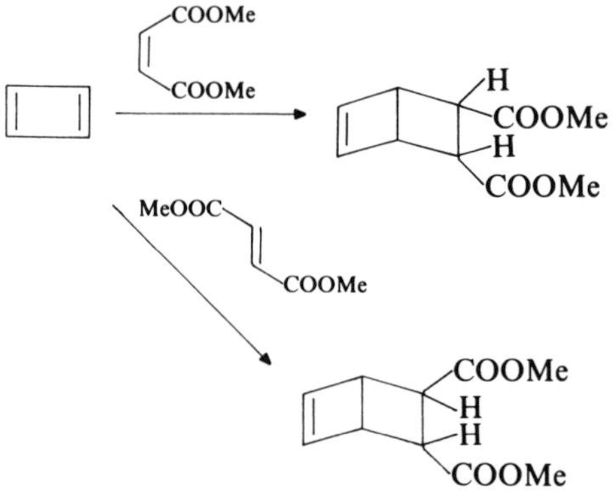

This indicates that the reactions involved were cyclo-addition reactions of a diene rather than of a diradical. A diradical species would result in the formation of a diradical intermediate which could then provide both the stereoisomeric products as follows:

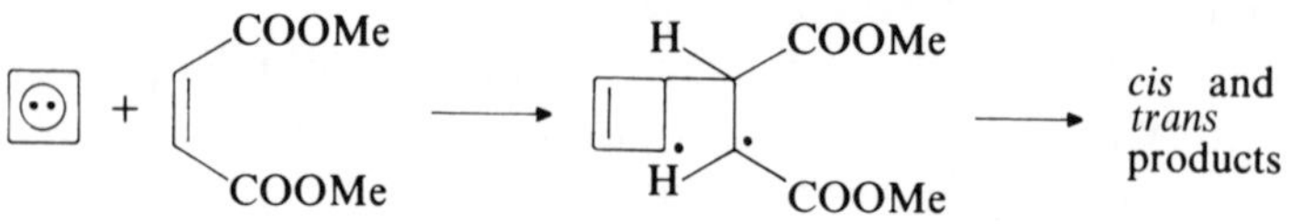

As further evidence, when 1,1-diphenylethylene was present in addition to either dimethyl maleate or dimethyl fumarate, the cyclobutadiene reacted exclusively with the esters and not with the alkene, which is a good radical trap. Additional more sophisticated evidence has also been gathered to show that cyclobutadiene behaves chemically as a diene rather than as a diradical.

Chemical Reactions of Cyclobutadienes

As seen above, cyclobutadiene reacts readily with dienophiles. Substituted cyclobutadienes, including highly substituted derivatives, react similarly with alkene and alkyne dienophiles, e.g.

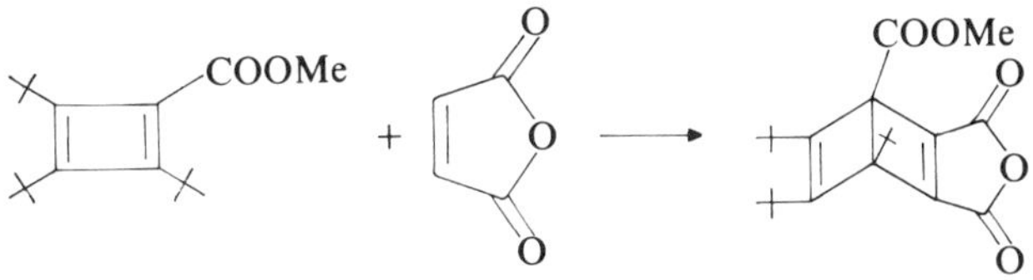

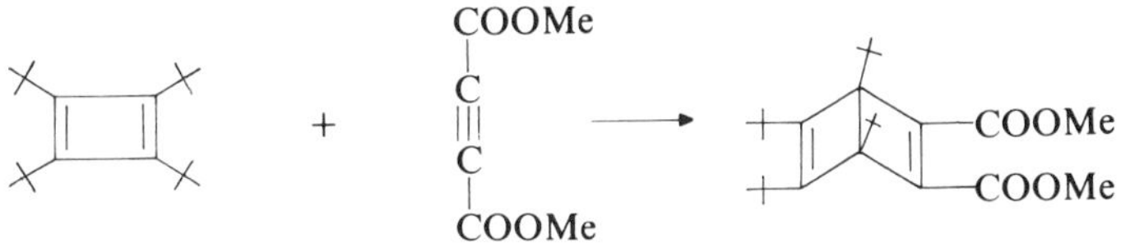

Reactions are, in general, stereospecific and may also be regiospecific, e.g.

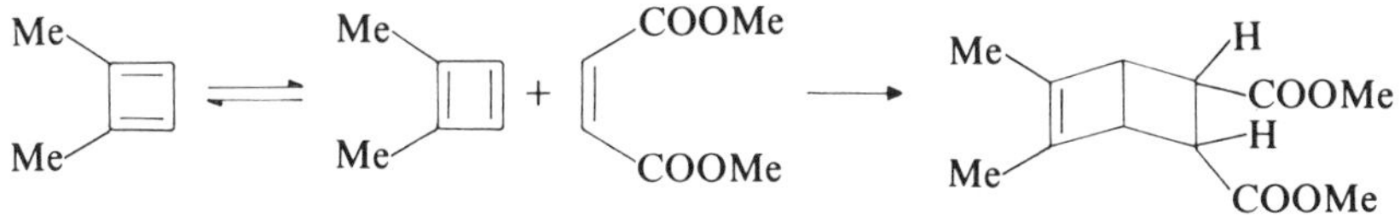

The equilibrating mixture of 1,2- and 1,4-diphenylcyclobutadiene reacts with only moderately reactive dienophiles, such as 1,4-benzoquinone, to give only products derived from the 1,4-diphenyl isomer. However, with more reactive dienophiles, such as tetracyano-ethylene, products from both the 1,2- and 1,4-diphenyl isomers are obtained:

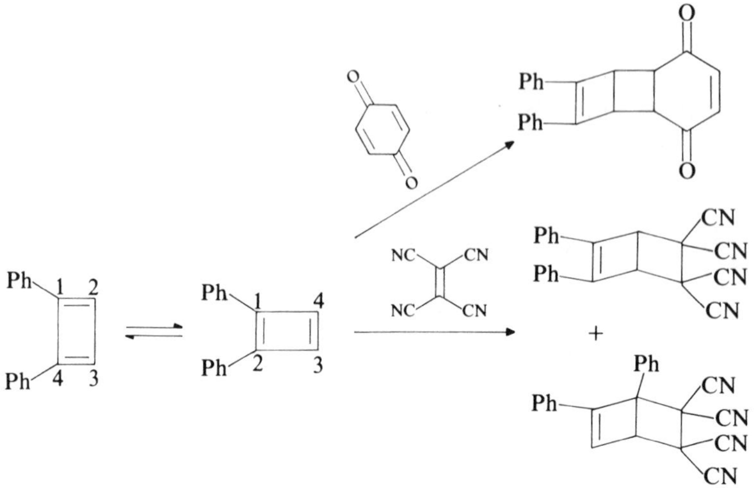

It is suggested that in the case of less reactive dienophiles equilibration between the 1,2- and 1,4-diphenyl isomers is fast compared to the cyclo-addition reaction, but in the case of more reactive dienophiles the rates of cyclo-addition and equilibration reactions are more nearly equal.

Cyclobutadiene can also react as a dienophile with other dienes, e.g.

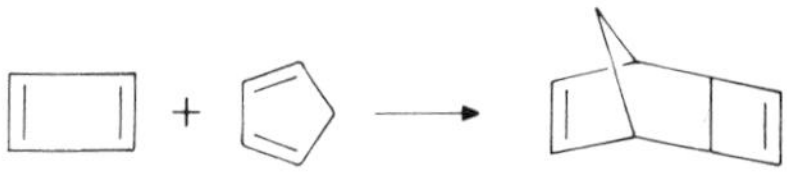

t-Butyl substituted cyclobutadienes have been shown to react not only as dienes or dienophiles but also with electrophiles, nucleophiles or radicals, e.g.

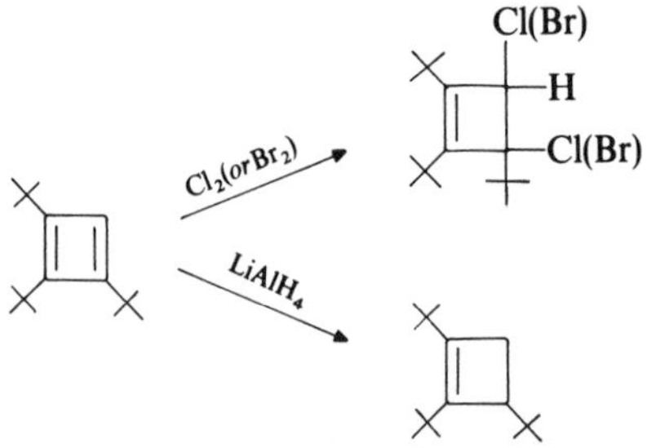

Their ability to behave as nucleophiles or electrophiles has been attributed to their having a low-lying LUMO (lowest unoccupied molecular orbital) and a high-energy HOMO (highest occupied molecular orbital). The autoxidation reactions mentioned earlier are examples of reactions involving radicals.

Cyclobutadienium and Cyclobutadienide Salts

Removal of two electrons from a molecule of cyclobutadiene should leave it with two electrons in a bonding orbital, and the resultant doubly charged cation meets the requirements of Hückel's rule, having two π-electrons delocalized over the ring.

First attempts to prepare such dications led instead to the formation of monocations, e.g.

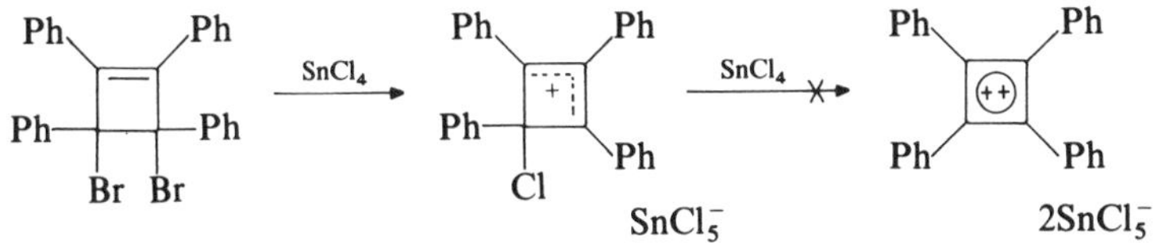

Use of a stronger acid is required to obtain a solution of the dication:

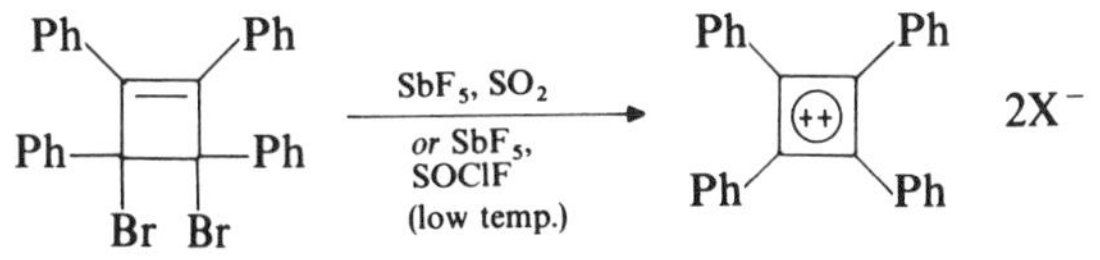

The tetramethyl analogue has been prepared similarly. Their structures are confirmed by their n.m.r. spectra; for example, in the case of the tetramethyl compound the four methyl groups provide a sharp singlet in the ^{1}H-n.m.r. spectrum.

Calculations indicate that the cyclobutadienium dications are probably non-planar, but this may not prevent there being strong stabilization from the two-π-electron system.

The cyclobutenium monocations are of interest in that they may be allylic cations, as shown above, or alternatively there may be 1,3-π-overlap between the ends of the allylic system, to provide what is a homocyclopropenium cation. (See Chapter 9 for cyclopropenium salts and Chapter 8 for a discussion of homoconjugation.) These two forms are represented as:

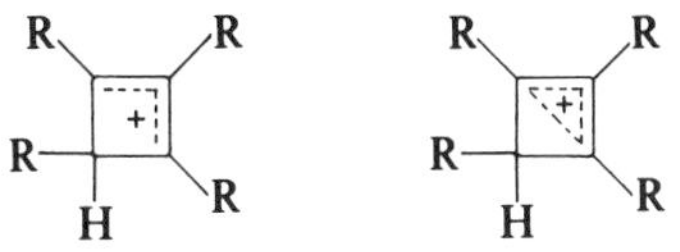

N.m.r. evidence suggests that when the substituent groups at the ends of the allylic system are phenyl there is only a very small contribution from 1,3-π-overlap, but that when there are methyl substituents this overlap is enhanced. In the case of the unsubstituted cation (R = H), its n.m.r. spectra indicate that overlap is strong, and it is truly an example of a homocyclopropenium cation. It was prepared in solution as follows:

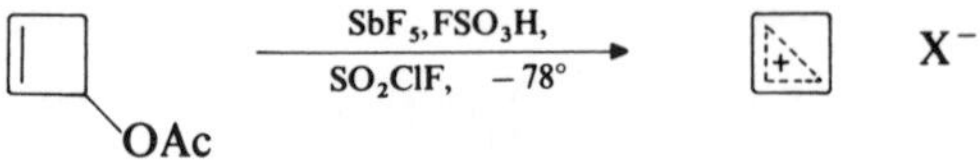

The n.m.r. spectra also show that the methylene group is out of the plane of the other three carbon atoms, and, except at very low temperatures, the folded ring is undergoing rapid inversion.

Addition of acids to tetra-t-butyltetrahedrane has provided crystalline cyclobutenium salts, e.g.

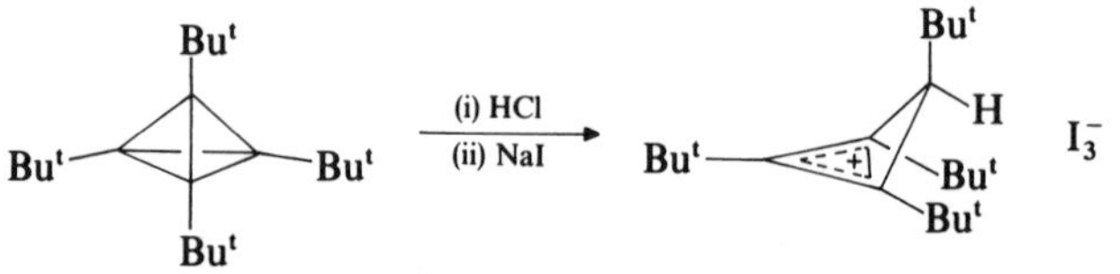

The crystal structure justifies a description of this salt as a homocyclopropenium salt; the ring is folded and the 1,3-distance is short, in accord with 1,3-π-overlap.

Addition of two electrons to cyclobutadiene to provide a dianion would provide a system of six π-electrons. However, this is not likely to provide any great stabilization since the orbitals of square cyclobutadiene include only one bonding orbital; the next highest orbitals are both non-bonding. In addition, it is likely that, for a multiply charged small ring, electron repulsion would outweigh any stabilization arising from an array of six π-electrons.

The unsubstituted dianion has not been prepared but substituted derivatives have been obtained in solution:

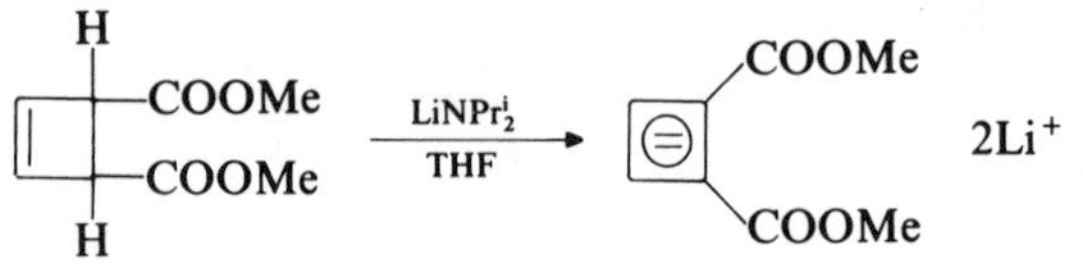

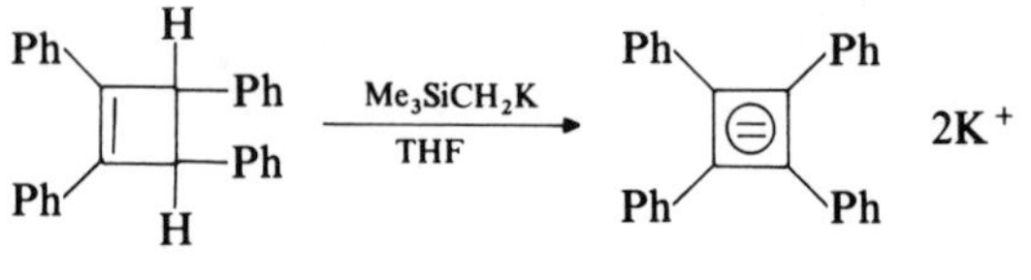

Addition of deuterium oxide to these solutions leads to the formation of dideuteriocyclobutenes. In both cases the n.m.r. spectra of the dianions indicate that the negative charge is largely located on the substituent groups rather than on the four-membered ring. pK studies suggest that there is no special stabilization of the cyclic dianions. It may be assumed that the unsubstituted dianion will be at least no more stable than these substituted derivatives, since it has no scope for any stabilization from delocalization of negative charge into substituents.

Azacyclobutadienes

2,3,4-Tri-t-butyl-azacyclobutadiene, or 2,3,4-tri-t-butylazete, has been prepared:

It was obtained as reddish-brown needles which were thermally exceedingly stable; it was unchanged after being heated at 100 °C in toluene for several days. It is also stable to photolysis but is sensitive to oxidation. N.m.r. spectra indicate rapid isomerism between the two possible valence isomers.

Annelation of Benzene Rings to Cyclobutadiene. Biphenylene

When attempts to prepare certain target molecules have failed and the task appears to be a difficult one, attempts have often been made to prepare instead benzo derivatives of the target compound, making the assumption that fusion of a benzene ring will provide a product which will be more easy to obtain and to handle.

This approach proved to be highly successful in the case of cyclobutadiene, and years before any evidence was obtained that cyclobutadiene had even a transient existence, its dibenzo derivative, biphenylene, had been isolated as pale yellow crystals:

Biphenylene is very different in character from cyclobutadiene in that it is stable, even at high temperatures, and can be kept for years without change. This change of character consequent upon the annelation of a benzene ring is now recognized as a common feature, and is discussed further in Chapter 11.

In general, when cyclic polymers are fused to give polycyclic compounds each constituent ring has an effect on its neighbours. As will be seen later, the π-electrons tend to arrange themselves so as to provide, first, as many separate cyclic arrays of *six* π-electrons as possible, and, second, as few arrays of $4n$ π-electrons as possible.

In the case of biphenylene, which has in all twelve π-electrons, this can most readily be achieved by having arrays of six π-electrons in each of the six-membered rings. X-ray crystallographic and electron diffraction studies show the bond lengths to be as follows (Å). (The figures quoted are those obtained by electron diffraction; the fact that there is good agreement between these and the X-ray results indicate that the molecules have similar structures in both the gaseous and solid states.)

These figures, coupled with calculations of bond orders based on these results, emphasize two points: (1) the low bond order of the bonds linking the six-membered rings, and (2) the extent to which bond lengths and bond orders alternate in the six-membered rings. This type of structure may be considered to result from the minimizing of cyclobutadienoid structure in the central ring. The partial fixation of the double bonds in the six-membered rings results in a lowering of the double-bond character of the bonds common to four- and six-membered rings. The twelve π-electrons are largely located in the six-membered rings. In addition to inhibiting any cyclobutadienoid character in the four-membered ring, the long low-order bonds linking the six-membered rings also minimize electronic interaction between the two six-π-electron systems in the outer rings which might lead to a peripheral destabilizing twelve-π-electron (i.e. $4n$, $n = 3$) system. There is some delocalization within the six-membered rings. In essence, there are two six-π-electron systems kept at arm's length by bonds of very low order.

Spectroscopic evidence shows that the two six-π-electron systems are not entirely independent of each other. The electronic spectrum shows absorption at much longer wavelengths than is the case for biphenyl, while the n.m.r. spectra are best interpreted in terms of a small paramagnetic ring current in the four-membered ring:

Preparation of Biphenylene

Three general methods have been used to prepare biphenylene and its derivatives; (1) removal of iodine (or bromine) from 2,2'-diiodo (or dibromo) biphenyls, (2) dimerization of benzynes, and (3) pyrolysis of benzo[c]cinnolines. Examples are:

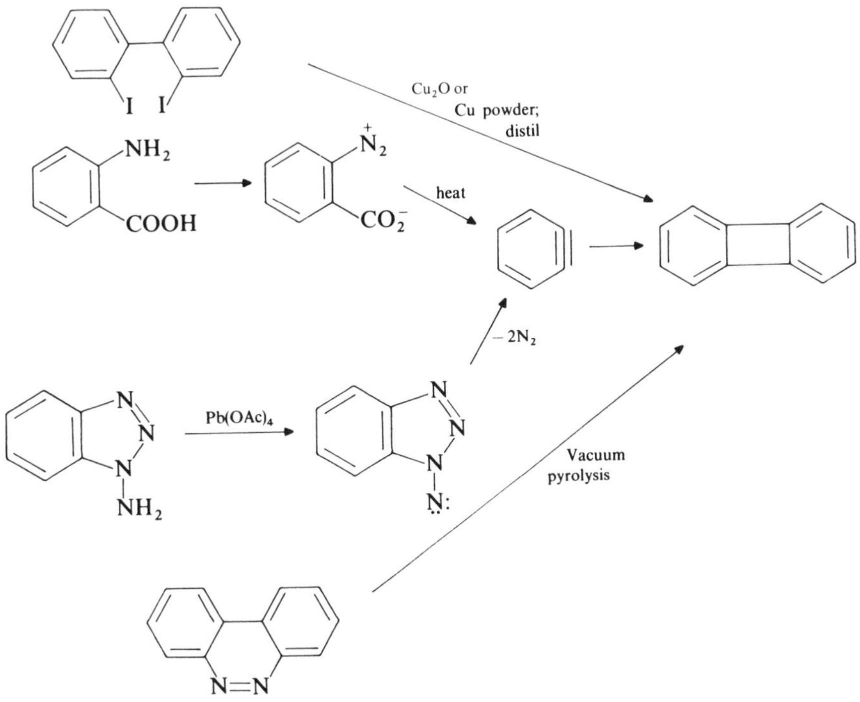

Reactions of Biphenylenes

As mentioned above, biphenylene is a very stable compound, which can be stored indefinitely under normal conditions. It is reduced catalytically to give biphenyl and oxidized to phthalic acid by chromic acid. It undergoes electrophilic substitution at the 2-position, e.g.

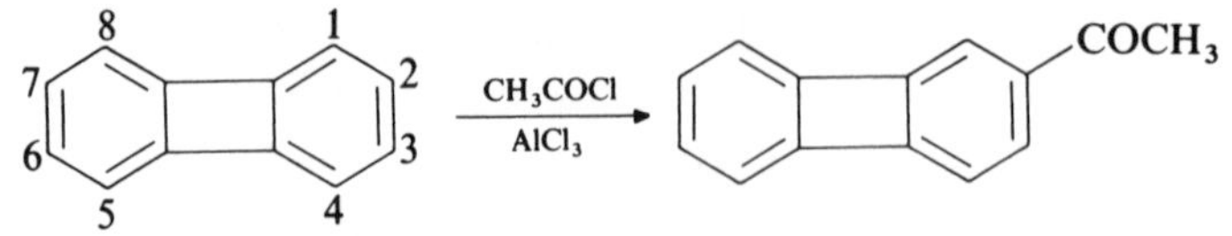

In some instances, electrophilic substitution is accompanied by ring-opening to provide benzocyclooctatetraene derivatives:

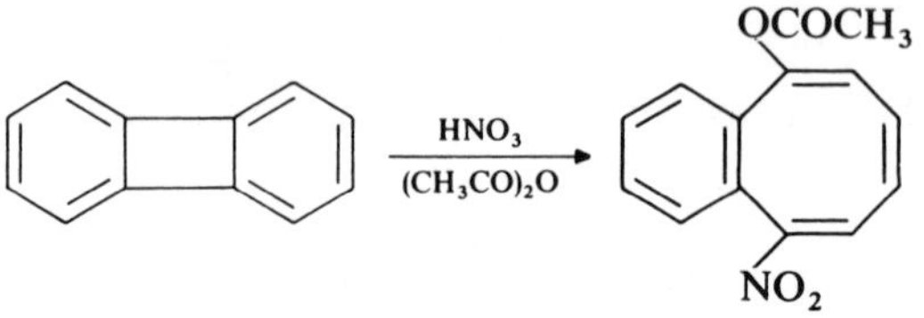

Halogenation frequently gives rise to complex mixtures of products, including halogenobenzocyclooctatetraene derivatives.

Not surprisingly, the presence of an electron-withdrawing substituent at the 2-position directs further electrophilic substitution into the other six-membered ring (mainly at the 6-position), whereas if an electron-donating substituent is present, further electrophilic substitution occurs in the same ring (at the 3-position).

Benzobiphenylenes

If a further benzene ring is annelated onto biphenylene two isomers are possible, linear or angular:

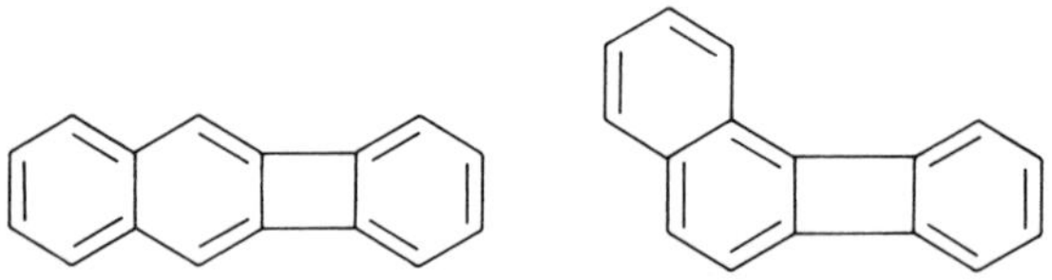

In effect, one of the fused benzene rings of biphenylene is replaced by a fused naphthalene ring. It is known that the 1,2-bond of naphthalene has more double-bond character than the 2,3-bond. Since it is desirable that the bond common to the four-membered ring and a fused six-membered ring should have as little double-bond character as possible, it follows that linear benzobiphenylene should be more stable than its angular isomer. Similar arguments apply to dibenzobiphenylenes and are in accord with the experimental findings.

In the bis(benzocyclobuteno)biphenylene illustrated below, the bonds in the central six-membered ring are extremely localized, with alternate bond lengths (averaged) of 1.335 and 1.495 Å:

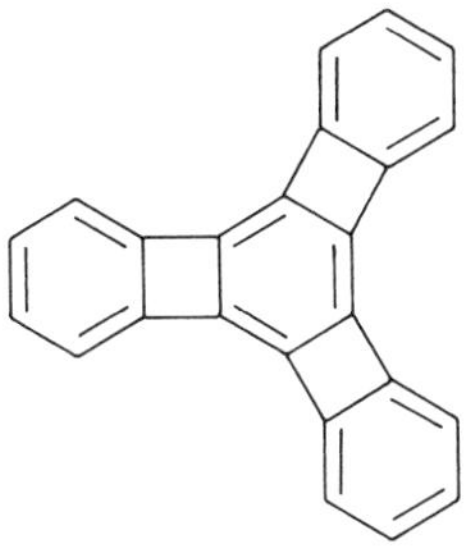

The thermal stability of this compound is surprising, in view of the amount of angle strain present in the molecules. It is stable neat up to 400 °C.

A number of heterocyclic azabiphenylenes (pyridinocyclobutadienes) have been prepared. Their structures resemble that of biphenylene. The dipyridinocyclobutadienes are hydrolysed by dilute aqueous alkali to give pyridinylpyridones, e.g.

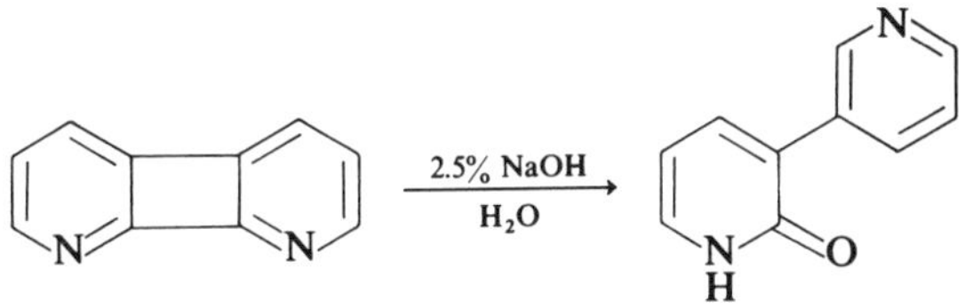

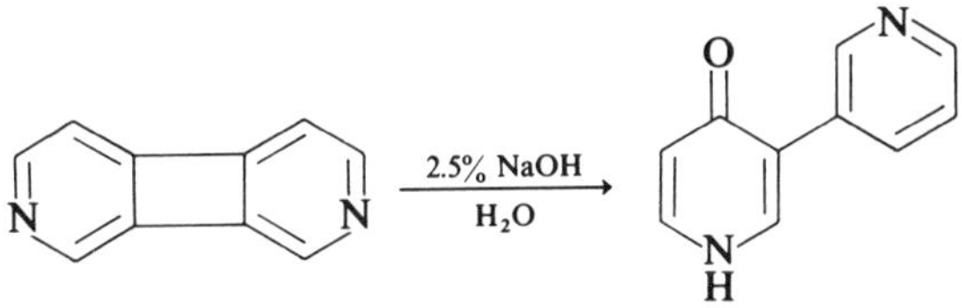

Both the benzothienocyclobutadienes shown below have been prepared:

(A)   (B)

In isomer (A) the bond joining the four-membered ring to the thiophen ring is a nominal single bond of the thiophen ring, whereas the corresponding shared bond in (B) is a nominal double bond. (For discussion of thiophen see Chapter 7.) Therefore it may be expected that (A) should be the more stable compound of the two, and this is the case. Compound (B) undergoes extensive decomposition in a few hours at room temperature whereas compound (A) is a crystalline solid which can be purified by vacuum sublimation.

Further Reading

For reviews on cyclobutadiene and its derivatives see G. Maier, *Angew. Chem.*, **100**, 317 (1988); *Angew. Chem. Int. Edn Engl.*, **27**, 309 (1988); T. Bally and S. Masamune, *Tetrahedron*, **36**, 343 (1980). A review concerned especially with annelated cyclobutadienes is by P. J. Garratt, *Topics in Nonbenzenoid Aromatic Chemistry*, Vol. I, John Wiley, New York, 1973, pp. 95ff.

Biphenylenes have been reviewed by J. W. Barton in *Non-benzenoid Aromatics*, Vol. I, Academic Press, New York, 1969, pp. 32ff.

Metal complexes of cyclobutadiene have been reviewed by P. M. Maitlis and K. W. Eberius in *Non-benzenoid Aromatics*, Vol. II, Academic Press, New York, 1971, pp. 360ff.

See also Chapter 5 of the author's 1984 book (for details see Preface).

5
[10]Annulenes and Related Compounds

[10]Annulene (cyclodecapentaene) has ten π-electrons and thus is a $(4n + 2)$ annulene, $n = 2$. Consequently there has been much interest in possible similarities between this compound and benzene. However, as mentioned earlier, account must be taken of other factors besides Hückel's rule. In the case of [10]annulene steric factors loom large.

Benzene takes up a planar hexagonal shape in which the preferred angles for trigonal carbon atoms (120°) are used and in which there is no crowding between non-bonded atoms. This ideal situation cannot be achieved for [10]annulene.

Possible shapes which the ring might take up are the all-*cis*, mono-*trans*, di-*trans* and tri-*trans* forms:

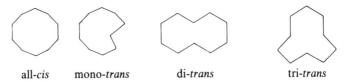

| all-*cis* | mono-*trans* | di-*trans* | tri-*trans* |

If the all-*cis* ring is planar the bond angles are considerably distorted from the normal trigonal angle; the internal angles are 144°. The di-*trans* form has bond angles of 120°, but a different problem arises in this case because it requires two hydrogen atoms to be situated within the ring, and to maintain the trigonal angles they should occupy the same space:

This could only be resolved by severe twisting of the ring, which poses further problems in that it would result in diminished overlap between the p orbitals of the carbon atoms. The tri-*trans* form suffers both from severe crowding of hydrogen atoms within the ring and from distorted bond angles.

The mono-*trans* form represents a compromise between the all-*cis* and di-*trans* structures and ameliorates to some extent the problems of the latter two forms. There is only one hydrogen atom positioned within the ring but there are still strained bond angles.

[10]Annulene has been prepared and shown to exist in both the all-*cis* and mono-*trans* forms. None of the di-*trans* or tri-*trans* forms have been detected. Both the all-*cis* and mono-*trans* forms try to escape from angle strain by becoming non-planar. Thus it appears that any energetic stabilization from delocalization of the π-electrons in a planar structure is insufficient to outweigh the steric disadvantages of planarity.

Ab initio calculations on all-*cis* [10]annulene confirm that energy is lowered substantially when the ring is made non-planar. When a planar geometry was considered, the energy surface in its vicinity was found to be very flat, and there was very little difference in energy between a delocalized structure and a structure with localized double bonds.

Preparation of [10]Annulenes

When *cis*-9,10-dihydronaphthalene is irradiated at $-60\,°C$, all-*cis*- and mono-*trans*-[10]annulene are formed. These have been separated as crystalline products by chromatography on alumina at $-80\,°C$:

Properties of [10]Annulenes

Spectroscopic evidence suggests that the all-*cis* isomer is non-planar. Its ^{1}H-n.m.r. spectrum has one singlet at $\delta 5.66$ and does not vary in the range $-40\,°C$ to $-160\,°C$. It is suggested that it exists as a mixture of different conformers. The energy barrier to interchange between these conformations is low and on the n.m.r. time scale an averaged structure, with the CH groups all equivalent to one another, is recorded.

The mono-*trans* isomer provides different ^{1}H-n.m.r. spectra at different temperatures. At $-40\,°C$ the spectrum consists of a sharp singlet at $\delta 5.86$. From $-40\,°C$ down to $-100\,°C$ the spectrum changes, consisting of two peaks at $-100\,°C$. No further changes occur below $-100\,°C$. It is suggested that, in addition to conformational changes, bond shifts take place; some flattening of the ring is involved. This can be represented as follows:

At $-40\,°C$ these changes are rapid on the n.m.r. time scale and only one averaged signal is seen.

At low temperatures $(-100\,°C)$, photochemical conversions between the all-*cis* and mono-*trans* forms proceed rapidly, the mono-*trans* being the favoured form. At somewhat higher temperatures, the all-*cis*- and mono-*trans*-[10]annulenes undergo thermal electrocyclization to give, respectively, *cis*- and *trans*-9,10-dihydronaphthalenes.

Catalytic hydrogenation at $-70\,°C$ of a solution containing the [10]annulenes provides cyclodecane in good yield, thus giving chemical evidence in support of the presence of the annulenes.

Bridged [10]Annulenes

The stable existence of di-*trans*-[10]annulene seems to be ruled out by the excessive crowding caused by the two hydrogen atoms which would perforce lie within the ring, unless the ring were severely twisted away from planarity. If, however, these two hydrogen atoms are replaced by one atom linked to the 1,6-positions of the ring this crowding would be eliminated:

Such compounds are known and are stable, for example, 1,6-methano[10]annulene:

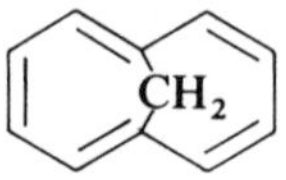

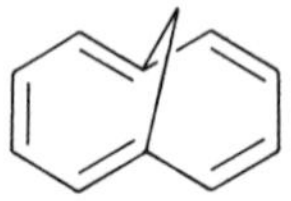

Related compounds are 1,6-oxido- and 1,6-amido[10]annulenes:

A tri-*trans*-[10]annulene would be even more crowded than its di-*trans* analogue, but again this crowding can be eliminated by bridging, and this is achieved in the methenoannulenes and cyclazines:

Tri-*trans*-[10]annulene   11-Methyl-1,4,7-   Cyclo[3.3.3.]azine
                          metheno[10]annulene

The methyl substituent on the central atom of the methenoannulene lies above the plane of the periphery and does not cause steric crowding.

A discussion of these examples of bridged [10]annulenes follows.

Preparation of 1,6-Methano[10]annulene

1,6-Methano[10]annulene is prepared from naphthalene by the following steps:

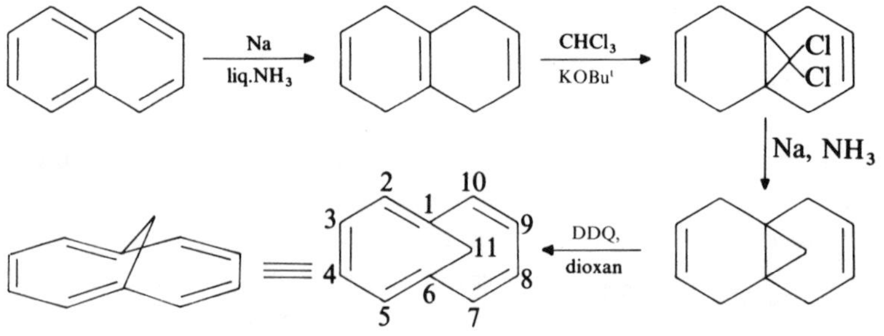

Birch reduction of naphthalene gives tetrahydronaphthalene. Dichlorocarbene adds to the inner double bond to provide a tricyclic structure. Reduction of the substituent chlorine atoms, followed by dehydrogenation, then gives the bridged annulene, possibly formed via a valence isomer which undergoes electrocyclic ring-opening of its three-membered ring:

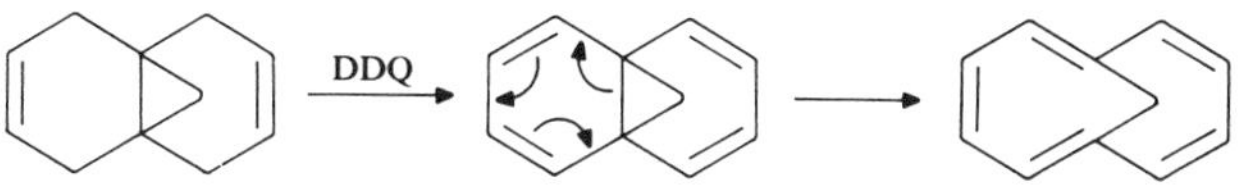

Structure

The electronic spectrum shows clearly the presence of an extended conjugated system. The ^{1}H-n.m.r. spectrum indicates a diamagnetic ring current, for the peripheral hydrogen atoms are deshielded ($\delta 7.2$, A_2B_2 system) and the bridging methylene group is strongly shielded ($\delta -0.5$, s). The spectra of 1,6-oxido- and 1,6-amido[10]annulenes indicate that they have similar structures.

An X-ray study shows that the periphery is not planar but that the 1- and 6-carbon atoms lie out of plane:

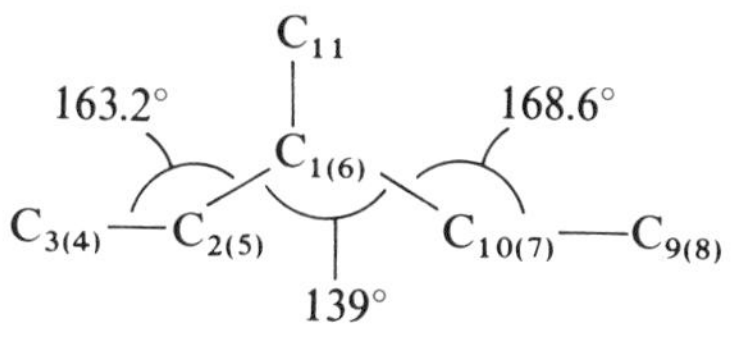

163.2° C$_{11}$ 168.6°
C$_{1(6)}$
C$_{3(4)}$—C$_{2(5)}$ C$_{10(7)}$—C$_{9(8)}$
139°

Bond angles

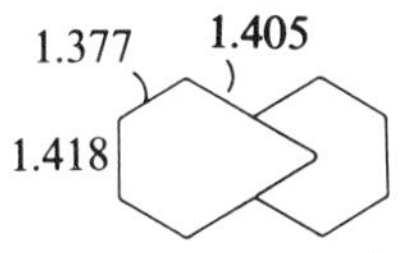

1.377   1.405
1.418

Bond lengths (Å)

Average bond lengths are as shown, and indicate that there is significant delocalization of π-electrons in the periphery, despite the lack of overall planarity. It is possible that overlap of the π-orbitals with consequent improvement of delocalization and stabilization of the system is enhanced by some modest rehybridization of the orbitals. The distance apart of the 1-C and 6-C atoms (2.235 Å) suggests that there is effectively no bonding between these atoms.

Certain substituent groups or atoms on the bridging carbon atom, notably methyl or cyano groups, result in differences in the carbon–carbon distances. There is more marked alternation of bond lengths in the periphery and the C(1) and C(6) distances are much shorter (~ 1.6–1.8 Å). It is thought that these compounds exist preferentially in a *norcaradiene* form:*

The bonds linking C(1) and C(6) are, however, unusually long.

Equilibration between annulene and norcaradiene forms proceeds via electrocyclic reactions involving six electrons. In order to investigate the situation further, the ^{13}C-n.m.r. spectra of a series of 11-substituted 1,6-methano[10]annulenes at different temperatures has been studied. Some of these compounds provide spectra which do not change with variation in temperature, and they appear to exist, within the limits of detection, exclusively as annulenes. However, the spectra of compounds having either bulky or electron-withdrawing substituents at C(11) change according to the temperature, indicating rapid exchange between annulene and norcaradiene forms. In some cases the annulene form, and in others the norcaradiene form, appears to be the more stable. The two valence tautomers are very close in energy and are separated by only a small energy barrier. These conclusions are supported by other spectroscopic information and by theoretical studies. For example, *ab initio* calculations suggest that the annulene form of 1,6-methano[10]annulene is more stable than the norcaradiene form by only 4.5 kcal mol^{-1}, and that the activation energies of interconversion are 7.3 and 2.3 kcal mol^{-1}. Thus the barrier separating annulene and norcaradiene forms is very small, and for derivatives of this system, depending on the substituents present, the product obtained experimentally might be either the more stable isomer or a fluxional mixture.

Two other complicating factors also need to be considered. One possibility is that the most stable form is an annulene, but with fixed double and single bonds rather than a delocalized structure. This appears to be the case for a tetra(trimethylsilyl) derivative, i.e.

*The possible bicyclic isomer of cycloheptatriene is known as *norcaradiene*:

hence the use of the term 'norcaradiene form' in the present context.

In the crystalline state this compound exists as a polyalkene. N.m.r. spectra show that this is also the case in solution at low temperatures but that at room temperature there is rapid valence tautomerism between the two different Kekulé forms.

The other possible complication is that of *homoconjugation* across the 1,6-positions of the ring. In Chapter 4 the possibility of 1,3-interactions across a cyclobutenium ring was mentioned. In homoconjugation π-electronic interaction takes place across a gap spanned by a methylene group. A classic example is the homotropylium ion, discussed in Chapter 8:

Homoconjugation of six π-electrons could be considered for 1,6-methano[10]annulene, as shown in the following formula:

There is spectroscopic and chemical evidence in support of some contribution from this type of structure, which could help to account for the observed alternation in peripheral bond lengths.

It is interesting to note that when the bridge is longer than one atom the bridged annulenes appear to exist entirely in a tricyclic form. Examples are:

and

In these molecules the bridging causes severe deformation of the peripheral ring and extreme steric hindrance to conjugation around the ring. In consequence, there is insufficient delocalization energy to favour the annulene structure and this, together with steric factors, leads to the tricyclic structure being preferred. Also, in these compounds homoconjugation is not possible.

The factors contributing to the detailed overall structures of 1,6-bridged [10]annulenes have been considered in some detail to emphasize the complexity of the situation. It is not sufficient to consider only the effects depending upon Hückel's rule and, in this and most other cases, it is necessary to take into account any other relevant factors. Having done this as well as possible, the outcome may be that it is not easy to decide between alternative possibilities because the energy differences between them may be small. In consequence they may be easily affected by external circumstances such as crystal forces in solids or solvation in solutions. Therefore, in some cases a compound may take up one structure in one environment but a different one in others.

Fortunately, for many compounds the complications are fewer and more decisive answers are possible, but in all cases care must be taken to recognize as far as possible all the contributing factors.

Chemical Properties of 1,6-Methano[10]annulene

Reflecting the complications of the structure of 1,6-methano[10]-annulene, its chemistry is also not entirely straightforward, showing resemblances to alkene and benzene chemistry. For example, like benzene, it undergoes electrophilic substitution reactions and has been brominated by bromine or N-bromosuccinimide, chlorinated, nitrated using various reagents, sulphonated, and acylated under Friedel–Crafts reaction conditions. In each case substitution takes place at the 2-position. However, in the case of bromination, reaction has been shown to proceed via a dibromo-adduct:

When dinitrogen peroxide was used as nitrating agent a similar intermediate adduct was identified.

1,6-Methano[10]annulene does not take part in a Diels–Alder reaction with maleic anhydride at 80 °C, but at higher temperatures it gives an adduct derived from the tricyclic valence isomer of the annulene. It also reacts with 1,2,4-triazoline-3,5-diones, and some of its 11-substituted derivatives react similarly with acetylenes:

Z = Me, Br, CN

1,6-Methano[10]annulene is a stable compound which does not polymerize and is stable to atmospheric oxygen, although it adds singlet oxygen. It has been reduced by lithium to a paramagnetic dianion; ^{1}H-n.m.r. signals appear at δ1.59 and 3.07 for the peripheral hydrogens and at δ11.64 for the bridging methylene group. This dianion appears to be stable at room temperature. Exposure to air leads to reformation of the parent annulene.

An 11-oxo derivative has been prepared:

This compound is remarkably stable, and requires temperatures above 200 °C to eliminate carbon monoxide. Spectroscopic evidence and an X-ray crystal structure provide no evidence for electronic interaction between the carbonyl group and the peripheral conjugated system.

1,6-Oxido- and 1,6-Amido[10]annulene

1,6-Oxido[10]annulene is stable to alkali but is converted by acids into a benzoxepin and/or α-naphthol. It has been nitrated, using copper(II) nitrate in acetic anhydride, but other attempted electrophilic substitution reactions have given instead naphthalene derivatives. For example, an attempted acetylation provided α-naphthol acetate. With bromine it reacted by addition, and it gives a cyclo-addition product with 1,2,4-triazoline-3,5-diones. The oxygen atom is readily removed by catalytic reduction, giving naphthalene, which is itself reduced further to tetralin.

1,6-Amido[10]annulene is protonated or acetylated on nitrogen. On sulphonation it produces α-naphthylaminesulphonic acids. It too gives a cyclo-addition product with 1,2,4-triazoline-3,5-dione.

1,5-Methano[10]annulene

The isomeric 1,5-methano[10]annulene has been prepared:

This is a stable compound, resistant to heat up to 300 °C. It undergoes cyclo-addition reactions with electron-deficient alkenes and alkynes.

There is spectroscopic evidence for homoconjugation across the 1,5-carbon atoms in some methoxy derivatives of this bridged annulene. Calculations suggest that for the 1,5-bridged isomer also there is only a small energy difference between localized and delocalized forms, steric misalignment of the $p\pi$ atomic orbitals probably diminishing the contribution of the delocalized form. As in the case of the 1,6-methano isomer, this misalignment may be somewhat alleviated by modest rehybridization of the orbitals.

1,4,7-Metheno[10]annulene

This 11-methylmetheno[10]annulene is a stable yellow oil whose ^{1}H-n.m.r. spectrum shows it to be diatropic, with signals at $\delta 7.53$–7.92 for the outer hydrogens and at $\delta -1.67$ for the methyl group.

In this case the 1-, 4- and 7-positions are sufficiently far apart to make transannular homoconjugation unlikely. Calculations suggest that this system is more likely than methano[10]annulene to resemble a planar [10]annulene. When allowance is made for electron correlation, a delocalized structure appears to be slightly lower in energy than a structure with localized single and double bonds.

An X-ray study of a dicarboxylic acid derivative shows that the molecules are dish-shaped, and that the central carbon atom has tetrahedral geometry. All the peripheral bonds have some double-bond character.

When this methenoannulene is heated in boiling xylene the methyl group migrates to a peripheral bridgehead site:

In this process the [10]annulene system is replaced by a benzene system, which is energetically favoured.

This methenoannulene undergoes electrophilic substitution, and has been nitrated, acetylated and alkylated. It is interesting that, in the case of the 11-benzyl analogue, electrophilic substitution takes place exclusively in the ten-π-system rather than on the benzene ring of the benzyl group.

11-Alkylmetheno[10]annulenes do not undergo cycloaddition reactions at room temperature, but do add to 4-phenyltriazoline-3,5-dione when the mixture is heated.

8-Hydroxy-11-methyl-1,4,7-metheno[10]annulene appears to exist in the hydroxy form; there is no spectroscopic evidence for the presence of any keto-tautomer. This was the first annulenol with a ring larger than six-membered to be isolated:

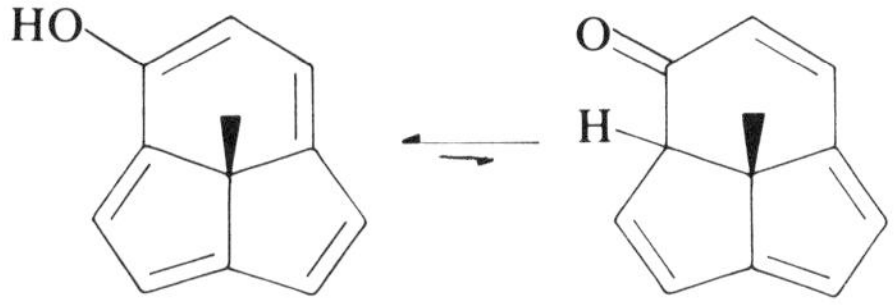

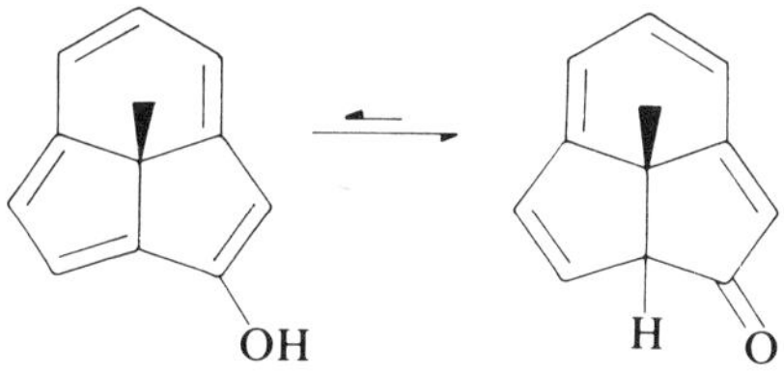

In contrast, the 2-hydroxy compound appears to exist entirely in the keto form. It is suggested that, in this case, any stabilization afforded by the annulenic structure in the enol form is insufficient to compensate for the increased steric strain in going from the keto to the enol form.

A nitrogen atom may take the place of the central CR group, giving a compound known as a *cyclazine*:

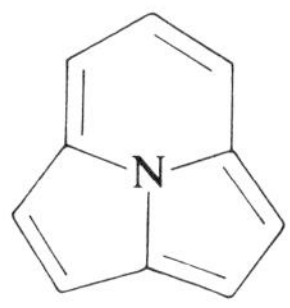

cyclo[3.2.2]azine

This is a very stable compound, to heat, light and air. It has a very weakly basic nitrogen atom; its ^{1}H-n.m.r. spectrum ($\delta 7.2-7.9$) indicates that the compound is diatropic. It undergoes electrophilic substitution, e.g. bromination, nitration and acetylation.

To regard cyclazines purely as bridged annulenes must, however, be oversimplistic, since, especially in planar cyclazines, electronic interactions with and through the central nitrogen atom, even if limited in extent, must be taken into consideration.

Aza[10]annulenes

Aza[10]annulene itself has not been described but a number of 2-aza- and 3-aza-1,6-methano[10]annulenes have been prepared:

The 2-aza compound is a stable yellow liquid whose electronic and n.m.r. spectra resemble those of 1,6-methano[10]annulene. The protons of the bridging methylene group provide a double doublet ($\delta 0.65$, -0.4, $J = 8.4\,\text{Hz}$).

In contrast, the 3-aza compound rapidly polymerizes, even in the absence of air, although from its spectra and an X-ray study it appears to have a delocalized π-electron system.

Further Reading

T. L. Burkoth and E. E. van Tamelen, *Non-benzenoid Aromatics*, Vol. I, Academic Press, New York, 1969, pp. 63ff.
S. Masamune and N. Darby, *Acc. Chem. Res.*, **5** 272 (1972).
See also Chapter 7 of the author's 1984 book (for details see Preface).

6

Larger Ring Annulenes

Interest in the chemistry of annulenes has largely concerned their resemblance or otherwise to benzene, and in particular the differences between those having $(4n + 2)$ and those with $4n$ π-electrons. In addition to their electronic structures, steric factors are of interest and importance, since these affect the shapes of the molecules, and may prevent their being sufficiently planar to allow effective delocalization of the π-electrons. Especially for the relatively smaller sized rings, a similar problem arises as with [10]annulene, in that if, in order to minimize angle strain, the molecules take up structures with *trans* double bonds, then some of the hydrogen atoms must be within the ring, and this results in intramolecular crowding. Various estimates have been made of the size of the ring required to permit uncrowded inner hydrogen atoms, ranging from [30]annulene downwards. It seems that the internal crowding in [18]annulene is sufficiently small to allow the molecules to take up an almost planar shape.

As mentioned in Chapter 2, it was suggested that as ring size increased, the stabilization of [4n + 2]annulenes and the destabilization of [4n]annulenes should decrease, and, for [4n + 2]annulenes as well as for [4n]annulenes, alternate double and single rather than hybridized bonds might be expected.

Annulenes are in general rather unstable compounds. Not much is known about their chemistry, largely because of their unavailability in sufficient quantity, although [16]annulene and [18]annulene have been studied more than the others. [18]Annulene does undergo some electrophilic substitution reactions, but, as described later, these reactions may not follow the usual mechanisms of electrophilic substitution.

A complicating factor is that for annulenes a variety of reaction pathways, for example, valence tautomerism to give bicyclic isomers, are available, which are not possible for benzene. This also makes it difficult to make direct comparisons of the chemistry of annulenes with that of benzene.

Preparation of Annulenes

Two general methods have been used for the preparation of annulenes, involving, respectively, ring closure of diynes and photolytic ring opening of polycyclic valence isomers of annulenes.

When α, ω-diynes are oxidized in aqueous ethanolic solution by oxygen in the presence of copper(I) chloride and ammonium chloride some cyclic dimer is produced, but the major part of the product consists of linear condensation products. Alternatively, reaction of α, ω-diynes with copper(II) acetate in pyridine provides a mixture of cyclic products of different ring sizes:

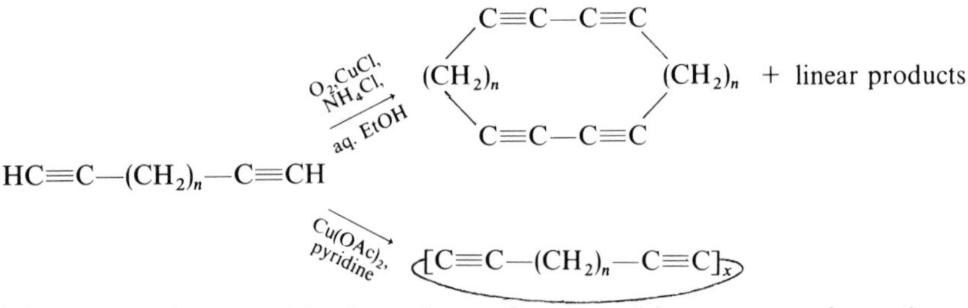

Many annulenes, with ring sizes of up to thirty atoms, have been prepared, using one or other of these methods of ring closure of diynes as the first step. The preparation of [18]annulene, as described in *Organic Syntheses*, **54**, 1 (1974), may be taken as an example:

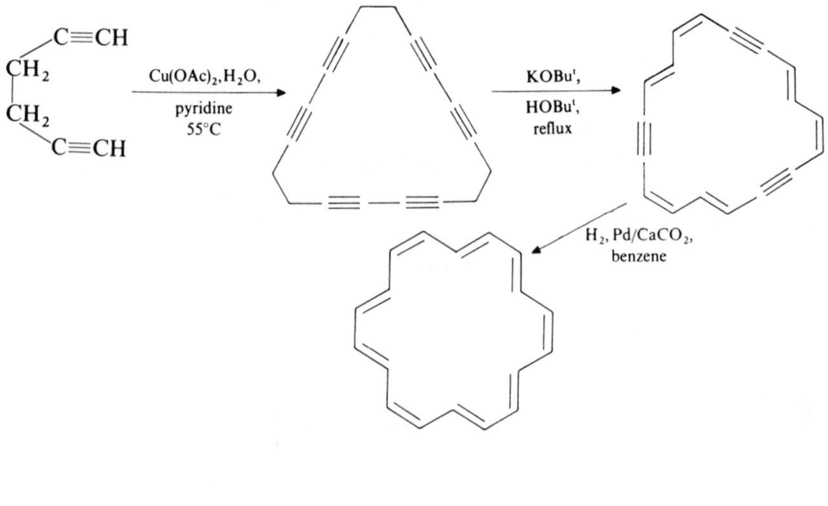

The initial reaction of the diyne gave a complex mixture, which was separated by chromatography an alumina to give the trimer shown (6%), together with a cyclic tetramer (6%), pentamer (6%) and hexamer (3%). When the trimer was treated with potassium t-butoxide prototropic rearrangements took place to give a completely conjugated polyenyne, or *dehydroannulene*. Partial hydrogenation of this compound over palladium gave crystalline [18]annulene. The cyclic tetramer and pentamer obtained in the first step were converted similarly into [24]annulene and [30]annulene, respectively. Other annulenes have been prepared, starting from other α, ω-diynes.

The conversion of a dimer of cyclooctatetraene into [16]annulene serves as an example of the preparation of annulenes from polycyclic valence isomers:

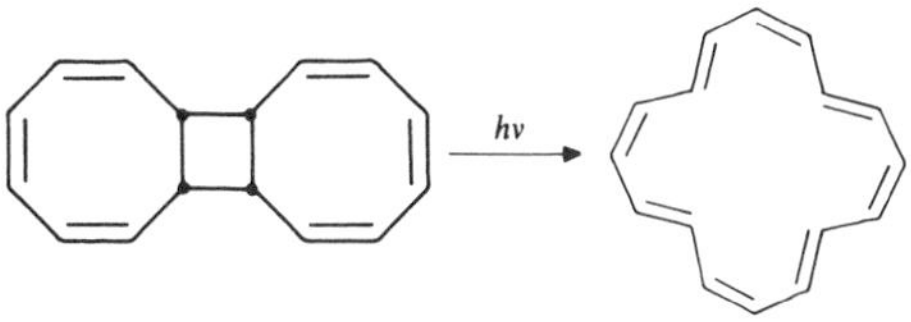

Properties of Annulenes

None of the annulenes is very stable, and all decompose in a short time. [18]Annulene is the most stable; its decomposition is accelerated by light. When heated in solution [18]annulene is converted into a mixture of benzene and a benzocyclooctatriene:

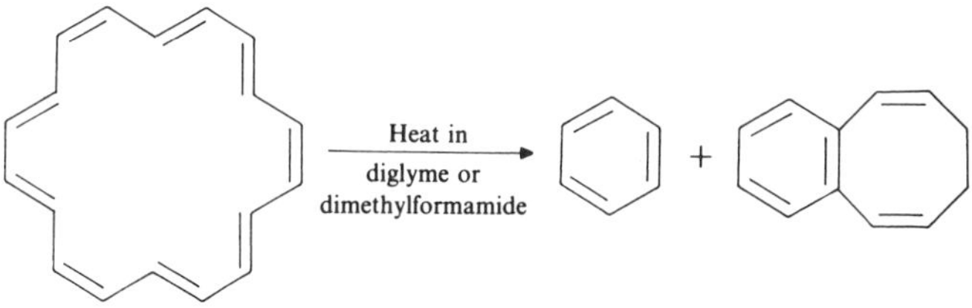

Reaction has been shown to proceed via the initial formation of a tetracyclic valence isomer. Valence isomerization is a common feature of annulenes. For example, [12]annulene is converted photolytically or thermally into bicyclo[6.4.0]dodecapentaenes:

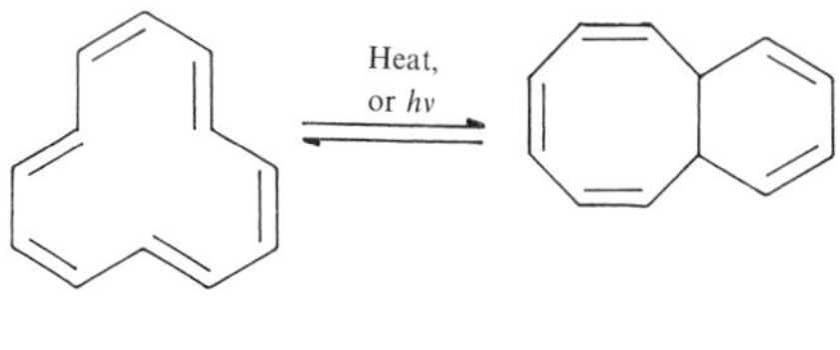

Annulenes are also sensitive to oxygen. [16]Annulene is very sensitive, but, for this reason, the crystals can be kept at least for several months, because a protective layer is formed on the surface of the crystals. For the same reason, freshly prepared [16]annulene is soluble in a range of organic solvents, but when aged it is insoluble unless the crystals are crushed.

The annulenes are intensely coloured and have a series of intense absorption maxima in their electronic spectra. For example, [18]annulene has its main maximum at 369 nm, with a very large extinction coefficient ($\varepsilon = 303\,000$). There appears to be an alternation in the wavelengths of the main absorption peaks, depending on whether the annulenes are $4n$ or $(4n + 2)$ π-electron compounds, e.g. λ_{max} [14], 317; [16], 285; [18], 369; [20], 284 nm.

The ^{1}H-n.m.r. spectra of annulenes are temperature dependent. At lower temperatures separate signals appear for protons outside the ring and protons within the ring, but at higher temperatures these signals coalesce and eventually give only one signal. For example, [18]annulene gives multiplets at $\delta 9.28$ and $\delta - 2.99$ at $-60\,°C$ but one singlet at $\delta 5.45$ at $110\,°C$. This is ascribed to conformational mobility at higher temperatures; the inner and outer protons exchange positions too quickly to provide separate signals and consequently give only an averaged signal. The most striking feature of ^{1}H-n.m.r. spectra of annulenes is the difference between the low-temperature spectra of [$4n$]annulenes and [$4n + 2$]annulenes. The latter are diatropic, and the outer protons are deshielded, as in benzene, whereas the protons within the ring are shielded. In contrast, [$4n$]annulenes are paratropic; the outer protons are shielded and the inner protons are deshielded. This is illustrated by the chemical shifts listed in the following table.

Annulene	δ, outer protons	δ, inner protons
[12]	5.91	7.86
[14]	7.6	0
[16]	5.40	10.43
[18]	9.28	−2.99
[20]	4.1 to 6.6	10.9 to 13.9
[22]	8.5 to 9.65	−0.4 to −1.2
[24]	4.73	11.2 to 12.9

The relatively small difference between the chemical shifts for the inner and outer protons of [12]annulene is probably associated with marked twisting of the bonds in the ring, which is caused by severe crowding of the three inner hydrogen atoms:

[12]annulene

The difference in chemical shift between the signals representing the inner and outer protons of an annulene, $\Delta\delta$, has been taken as a measure of the diatropicity or paratropicity of the annulene, i.e. of the ring current induced in the molecules.

[12]Annulene and [16]annulene are both reduced by lithium to give dianions. There is a striking reversal in the ^{1}H-n.m.r. spectra of these anions. In contrast to the neutral annulenes, the outer protons of the dianions are deshielded and the inner protons are shielded.

Annulene dianion	δ, outer protons	δ, inner protons
$[12]^=$	6.23, 6.98	-4.6
$[16]^=$	7.45, 8.83	-8.17

These dianions have 14 and 18 π-electrons, respectively, and are thus $[4n + 2]$annulenes. This reversal of character when two electrons are added to annulenes provides outstanding evidence for the different types of spectra associated with species having, on the one hand, $(4n + 2)$ π-electrons and, on the other, $4n$ π-electrons spread over a ring. In accord with this, reduction of [18]annulene gives a paratropic dianion, with signals for the outer protons appearing at $\delta - 1.13$ and for the inner protons at $\delta 28.1$, 29.5.

Surprisingly, treatment of [16]annulene with fluorosulphonic acid at low temperature results in the formation of the [16]annulenium dication rather than a protonated species. This cation has a 14 π-electron system and provides a spectrum typical of a diatropic species. It is also noteworthy that the $[4n + 2]$ π-electron dianions derived from [12]annulene and [16]annulene are more thermally stable than the neutral annulenes, whereas the $[4n]$ π-electron dianion obtained from [18]annulene is less stable.

Of the annulenes, only the reactions of [18]annulene have been investigated to any extent. It undergoes addition reactions with bromine and with maleic anhydride. Thus its simple chemistry differs markedly from that of benzene. Under special conditions it undergoes reactions which provide products typical of electrophilic substitution, e.g.

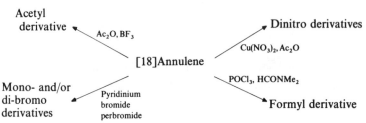

Acetyl derivative ← Ac_2O, BF_3 — [18]Annulene

Dinitro derivatives ← $Cu(NO_3)_2, Ac_2O$

Mono- and/or di-bromo derivatives ← Pyridinium bromide perbromide

$POCl_3, HCONMe_2$ → Formyl derivative

These reactions may not, however, proceed by straightforward electrophilic substitution mechanisms. [14]Annulene also undergoes an addition reaction with maleic anhydride.

X-ray crystallographic studies show that the ring of [18]annulene is almost planar in the crystal state, but that those of [14]annulene and [16]annulene are non-planar. [16]Annulene has alternate single and double bonds of length 1.46 and 1.34 Å. In [14]annulene and [18]annulene bond lengths vary, from 1.35 to 1.41 Å in the former and from 1.38 to 1.42 Å in the latter, but there is no obvious bond alternation as in [16]annulene.

Dehydroannulenes

Dehydroannulenes are fully conjugated cyclic polyenynes. An example was mentioned above as an intermediate in the synthesis of [18]annulene. The prefix *dehydro* is used to signify the absence of hydrogen atoms at the ends of a bond, i.e. the presence of a formal triple bond. Their naming is best illustrated by examples:

Monodehydro[14]
annulene

1,5-Bisdehydro[12]
annulene

1,5,9-Trisdehydro[12]
annulene

In a dehydroannulene there is a continuous conjugated system of
π-electrons. The 'extra' electrons of the triple bonds are orthogonal
to the conjugated system. Benzyne is an example of a dehydroannulene, being monodehydro[6]annulene.

The introduction of triple bonds in place of double bonds might
have a number of effects on the stability and bond structure of the
ring. First, it increases the rigidity of the ring. This could have a
stabilizing or destabilizing effect, depending on the geometry of the
ring in question. If the triple bond holds the ring in an unstrained
conformation, stability is increased, but if it results in the introduction
of angle strain, stability is lowered. Second, crowding of hydrogen
atoms is reduced because there are fewer of them. This is a generally
stabilizing influence. Third, the presence of the shorter triple bond
may induce bond alternation.

This last effect may become particularly important with larger rings
where theory in any case predicts a trend to bond alternation for all
annulenes. Some relief from crowding is likely to be important for
smaller rings, particularly where the crowding due to intramolecular
hydrogens causes distortion of the preferred geometry.

Whereas a system of alternate single and double bonds provides
two identical Kekulé forms, as in the case of benzene, this is not so
for dehydroannulenes. For example, shift of the π-electrons around
the ring of 1,7,13-trisdehydro[18]annulene (a precursor in the
synthesis of [18]annulene; see earlier) provides a very different form:

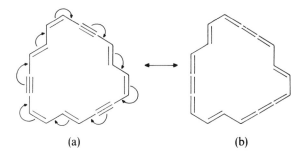

(a)   (b)

Although both forms contribute to the overall structure, (a) will
probably contribute more than (b).

Attempts to write canonical forms for dehydroannulenes inevitably
lead to allene structures, but there are examples where two equivalent
Kekulé forms can be written, each of which include both triple bonds
and allenic systems. An example is 1,8-bisdehydro[14]annulene:

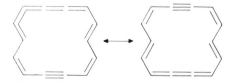

Note that the 'extra' π-electrons of a triple bond and the π-electrons forming the central double bond of the allenic system are both orthogonal to the main peripheral conjugated system. These two forms should contribute equally to the overall structure, and the molecule is best represented as follows:

X-ray structures are in accord with this representation. The molecule is centrosymmetric. The two potential triple bonds are 1.208 Å in length; all the other bonds are of lengths 1.378–1.403 Å, appropriate to a fully delocalized system of π-electrons. X-ray studies on a related bisdehydro[18]annulene gave a very similar picture, but for a bisdehydro[22]annulene there seemed to be evidence for some bond localization, in accord with ideas of decreased delocalization in larger rings.

However, dehydroannulenes appear to sustain diamagnetic or paramagnetic ring currents for ring sizes of at least up to thirty members. There is evidence that for dehydroannulenes with similar geometry, $\Delta\delta$, the difference between the chemical shifts for the inner and outer protons, decreases with increasing ring size.

Chemistry of Dehydroannulenes

It is difficult to generalize about the stability of dehydroannulenes. Some are more stable than the related annulenes, others are less stable. Too many factors come into play to permit simple explanations.

One very stable example is the 1,8-bisdehydro[14]annulene discussed above. This compound remained unchanged after exposure to air and light for a month. In its chemistry it more closely resembles benzene than do any of the annulenes. It does not add to maleic anhydride but does undergo electrophilic substitution reactions such as sulphonation, nitration and, albeit in low yield, acetylation.

1,5-Bisdehydro- and 1,5,9-trisdehydro[12]annulenes and an octa-dehydro[24]annulene have been reduced by alkali metals to give radical anions and dianions. These dianions have $(4n + 2)$ π-electron systems and are diatropic. Hence, as with annulenes, there is a dramatic reversal in their ^{1}H-n.m.r. spectra compared to those of the related neutral dehydroannulenes. For example, in the case of the bisdehydro[12]annulene the signal for the protons located within the ring shifts from $\delta 17.6$ to $\delta - 6.88$ on reduction.

Bisdehydro[18]-, -[22]- and -[26]annulenes have been reduced not only to dianions but to tetra-anions. Thus the diatropic neutral compounds are reduced in turn to paratropic dianions and diatropic tetra-anions, thus providing striking examples of the dependence of the tropicity of a ring system upon the number of π-electrons associated with it. In each case the diatropic ring-current effects are greater in the tetra-anions than in the corresponding neutral dehydroannulenes. The tendency of excess charge to be uniformly distributed over the entire π-electron system strongly reinforces the delocalization of the π-bonds. For either the neutral, dianionic or tetra-anionic species, the ring-current effects, diamagnetic or paramagnetic, decrease with increase of ring size.

When tetra-anions are formed from dianions, the production of a $(4n + 2)$ π-electron system will assist the process, but it will be opposed by electron–electron repulsion effects. In this connection it is notable that the corresponding bisdehydro[14]annulene could be reduced only to a dianion and not to a tetra-anion. This suggests that a minimum ring size is necessary to overcome the electron–electron repulsion involved in a tetra-anion, and that below this size a tetra-anion is, in consequence, not obtained.

Bridged Annulenes

As in the case of bridged [10]annulenes, a number of bicyclic and polycyclic compounds have been described which have a completely conjugated system around their periphery, which is bridged by saturated groups. Bridging could either stabilize or destabilize such a molecule, depending on whether it served to keep the periphery planar or to distort it away from planarity. In the bridged

[10]annulenes discussed in Chapter 5 there is a decrease of crowding compared with the (theoretical) parent annulenes, in that the inner hydrogen atoms of the latter, which would crowd each other, have been replaced by a single bonded atom.

For convenience, bridged annulenes will be separated into compounds with methano, ethano and metheno bridges, and each type will be considered separately.

Methano-bridged Annulenes

1,7-Methano[12]annulene forms brown-black needles and is paratropic:

The ^{1}H- and ^{13}C-n.m.r. spectra are temperature dependent. The signals for the bridging CH_2 group do not change but those associated with the periphery do. The changes are ascribed to equilibration between two different Kekulé-type valence isomers. The activation energy for this isomerization is small ($\leqslant 5$kcal mol^{-1}); it represents the difference in energy between the localized and delocalized structures for this compound, since interconversion between the two forms with alternate single and double bonds must presumably pass through the delocalized form. An X-ray structure of the symmetric 4,10-dibromo derivative indicates an alternation of single and double bonds around the periphery; the molecule is twisted.

Reduction of this bridged annulene electrochemically or by alkali metals provides solutions of its dianion, which is stable at room temperature and diatropic. The annulene ring is flattened compared to the neutral compound.

A series of 1,6-methanoannulenes has been prepared, with peripheral [12], [18], [20], [22], [24], [26], [28], [30], [32], [34] and [38] rings:

The [18] and [22] rings are diatropic and the [12], [20] and [24] rings are paratropic. The larger rings are atropic. As the size of the rings increases, wavelengths of the main absorption peaks in the electronic spectra alternate between the annulenes with $4n$ or $(4n + 2)$ π-electrons, as occurs with unbridged annulenes.

A related series of 1,6-methanobisdehydroannulenes with peripheral [26], [28], [30], [32], [34] and [38] rings has also been prepared:

Of this series, the [26], [30] and [34] dehydroannulenes are diatropic, the [28] ring is paratropic, and the [32] and [38] rings are atropic. Both the diatropicity and paratropicity decrease with an increase in ring size and appear to disappear at [38] and [32] for the $(4n + 2)$ and $4n$ systems, respectively. All these methanobisdehydroannulenes are more thermally stable than the corresponding monocyclic dehydroannulenes or annulenes, presumably because of the greater rigidity of the bridged compounds.

Polymethanoannulenes and Related Compounds

Examples of dimethanoannulenes are the two isomeric 1,6;8,13-dimethano[14]annulenes:

syn   *anti*

The two bridges may be *syn* or *anti* to each other.

There is a marked difference between the properties of these two compounds. The *syn* isomer forms stable orange crystals but the *anti* isomer is sensitive to oxygen. Electronic and n.m.r. spectra show that the *syn* isomer has a delocalized π-electron system around its periphery, whereas the *anti* isomer appears to be a mixture of two valence isomers with alternate single and double bonds, corresponding to the two Kekulé forms. N.m.r. spectra at different temperatures indicate that these two Kekulé forms are interconverting. This must presumably involve a delocalized transition state, which implies that the delocalized structure is of higher energy than the structure with localized double bonds.

The difference between the *syn* and *anti* isomers appears to be associated with the shapes of the peripheries. In the case of the *syn* isomer, models show that the periphery can be approximately planar, but in the *anti* isomer considerable distortion of the periphery from planarity is inevitable, which in turn inhibits delocalization of the π-electrons.

Thermally stable *syn*-1,6;8,13-bridged [14]annulenes having two oxido (—O—) bridges, one oxido and one methano bridge, and one amido and one methano bridge have been prepared, but their *anti*-isomers have not. There is a similar contrast between two isomeric dicarbonyl bridged [14] annulenes:

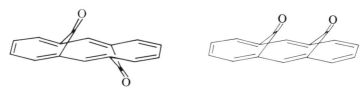

Either of these compounds might be expected to decompose readily to give the stable products carbon monoxide and anthracene. However, the *syn* isomer is very stable and resists flash vacuum pyrolysis at 500 °C. X-ray structures show that the periphery is virtually planar, despite repulsions between the two adjacent carbonyl groups. On the other hand, crystals of the *anti* isomer are sensitive to light. The *anti* isomer also polymerizes readily in air.

The great difference in properties between these *syn* and *anti* isomers provides impressive evidence of the effects of molecular geometry on the electronic structure of cyclic conjugated molecules. Some of the *syn* isomers have noticeably bent peripheries but none the less have delocalized π-electron systems. The *anti* isomers have strongly puckered rings which prevent full peripheral conjugation, and in consequence the molecules exist as polyalkenes, which can, however, interconvert between the two Kekulé forms. There is other evidence that [4n + 2]annulenes can tolerate considerable deformation of the ring skeleton from planarity without substantial weakening of the delocalization, provided that the torsional angles of the C—C bonds in the ring are not greater than about 45–55°.

Other dimethano[14]annulenes are also known, e.g.

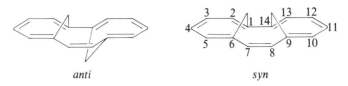

anti   syn

Spectra, and also an X-ray study of a substituted derivative, indicate that the *anti* isomer is a polyalkene with alternate single and double bonds. The *syn* isomer is diatropic, but the ring current appears to be weaker than in *syn*-1,6;8,13-dimethano[14]annulene. An X-ray examination indicates a small alternation of bond lengths over the end portions of the periphery [C(1)–C(6) and C(9)–C(14)], but the bond lengths in the central portion of the periphery are more appropriate to those found in a polyalkene. Crowding between the two methano bridges may be responsible, and it is possible that the best representation of this molecule is as depicted below, where there are two homoconjugated systems, each concerning six π-electrons, linked across C(6) to C(9) by an alkene bridge:

A number of these polymethanoannulenes have been reduced by lithium to give deeply coloured dianions, which are paratropic. For example, the dianion of *syn*-1,6;8,13-dimethano[14]annulene provides ^{1}H-n.m.r. signals for the peripheral hydrogens at δ2.70–3.56 and for the methano groups at δ8.50 and 11.95. Dications have also been made by reactions of polymethanoannulenes with SbF_5/FSO_2Cl they are also paratropic.

A different kind of oxido-bridged annulene is exemplified by 1,4;7,10;13,16-trioxido[18]annulene:

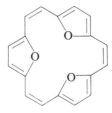

This compound is diatropic and has an electronic spectrum closely resembling that of [18]annulene.

Models suggest that it is planar, but that if the oxygen atoms are replaced by sulphur this is no longer so, because of the size of the sulphur atoms. In accord with this, n.m.r. spectra of the trithioxido analogue give no evidence for any appreciable ring current, and its electronic spectrum indicates that it is best represented as three thiophen rings linked by alkene-like bridges.

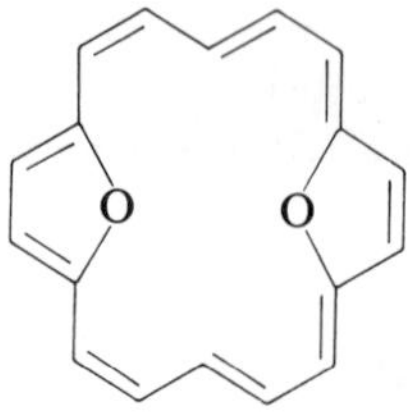

The dioxido[18]annulene above could, at the time of writing, go into *The Guinness Book of Records* as the neutral annulene showing the greatest observed difference in chemical shift between its inner and outer protons ($\delta - 5.89$ and $+ 9.13-10.00$). This is probably because the oxygen atoms serve to keep the ring firmly planar in a centrosymmetric conformation.

Ethano-bridged Annulenes

If the bridging carbon atoms of a bismethanoannulene are bonded to one another an ethano-bridged annulene results, e.g.

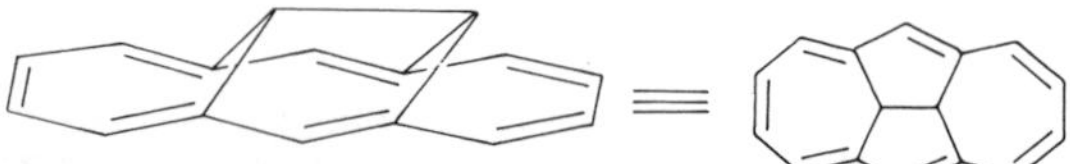

This compound is diatropic.

The most common examples of ethano-bridged annulenes are the *dihydropyrenes*:

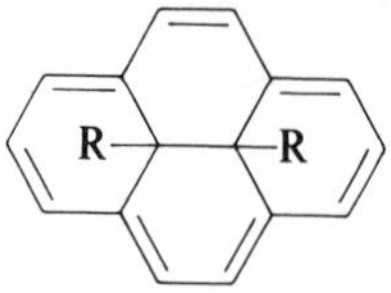

The most studied of these compounds have been the dimethyl derivatives, R = Me.

For all the dihydropyrenes, *cis* and *trans* isomers are possible:

Whereas the *trans* dimethyl isomer is a stable compound, the *cis* isomer reacts slowly with air and must be stored *in vacuo* in the dark. This is possibly due to there being less steric hindrance to attack by oxygen in the *cis* isomer, where attack can take place from the side where there are no protective substituent groups. In keeping with this idea, the parent *trans* compound, with R = H, is also susceptible to light and oxygen.

The dihydropyrenes are strongly diatropic, with deshielded peripheral protons and strongly shielded bridges, e.g. R = Me, CH_3, $\delta - 4.25$; R = H, inner H, $\delta - 5.49$.

The *trans*-dimethyl derivative is stable up to about 200 °C, but above that temperature migration of one of the methyl groups takes place:

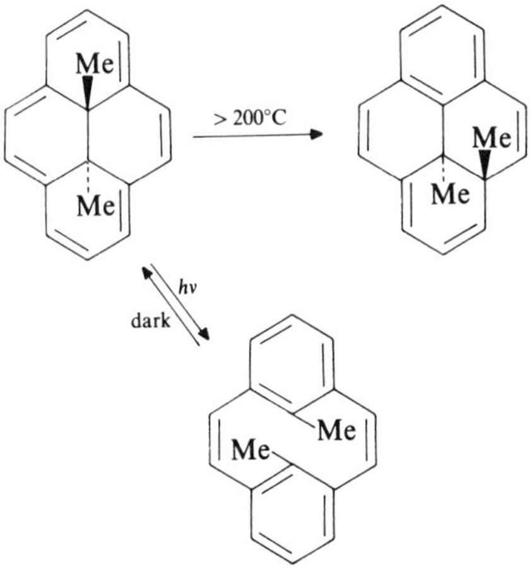

Light breaks the central carbon–carbon bond, but in the dark the process is reversed.

This dihydropyrene undergoes electrophilic substitution more readily than does benzene. It is easily deuteriated, nitrated, brominated, and acylated or alkylated in Friedel–Crafts reactions. It does not react with maleic anhydride.

An isomer of this dihydropyrene has also been prepared:

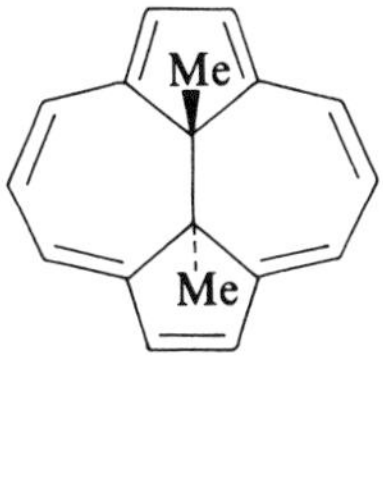

It, too, is stable and diatropic, and can readily be nitrated.

Both this latter compound and *trans*-dihydropyrenes can be reduced by potassium to give dianions which are strongly paratropic.

A rather more complicated bridged annulene is hexahydrocoronene (in coronene itself there are no saturated carbon atoms):

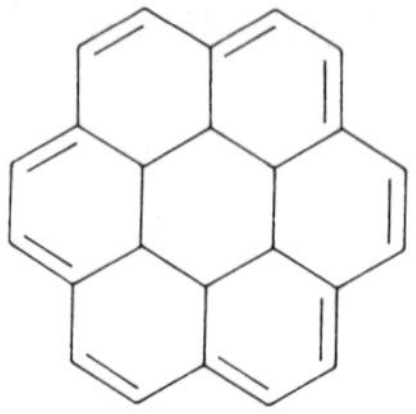

Its annulenic character is evident from its ^{1}H-n.m.r. spectrum, which has signals for the outer and inner hydrogens at, respectively, δ 9.30 to 9.55 and δ -6.5 to -8.0.*

Methenoannulenes

Metheno-bridged [12]- and [14]annulenes have been described:

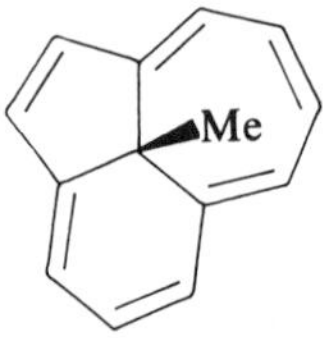

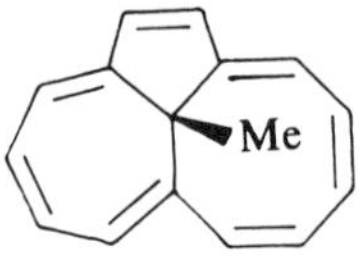

The blue metheno[12]annulene is stable up to 80 °C, and is paratropic. It is a polyalkene and appears to exist only in the one Kekulé form shown. The metheno[14]annulene is diatropic but appears to have some polyalkene character; it is likely that the eight-membered ring deviates from planarity, inducing π-bond alternation.

Cyclazines with twelve-π-electron peripheries are also known:

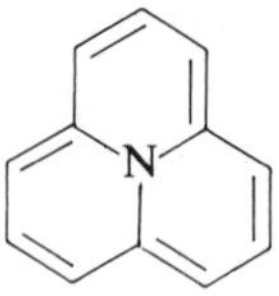

cyclo[3.3.3]azine

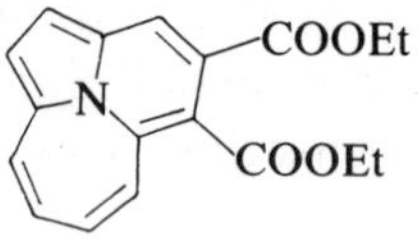

Diethyl [2.3.4]cyclazine-4, 5-
dicarboxylate

---

*Although the difference in chemical shift between the inner and outer hydrogens is even greater here than in the dioxido[18]annulene mentioned above, the compounds are not strictly comparable, since in the present case the inner hydrogen atoms are not directly attached to the annulene ring.

Cyclo[3.3.3]azine is stable as a solid under nitrogen but decomposes in a few minutes in air or in solution. It is paratropic. Only derivatives of the isomeric cyclo[2.3.4]azine system have been described (for example, the compound shown above). It, too, is paratropic, but the shielding of the protons is different in the three rings, and the n.m.r. spectrum of this compound is best interpreted if the molecule is considered to have characteristics both of a cyclazine with peripheral conjugation and of an indolizine with a butadiene bridge. Other cyclo[2.3.4]azines are similar.

Aza-annulenes and Bridged Aza-annulenes

Aza[14]annulene and aza[18]annulene are strongly coloured compounds which are diatropic, thus resembling their carbocyclic analogues. With dry hydrogen chloride aza[18]annulene is protonated on nitrogen giving a diatropic cation which is stable in air and can be reconverted into the aza-annulene by the action of ammonia.

A number of 2-alkoxyazadehydroannulenes have been prepared, e.g.

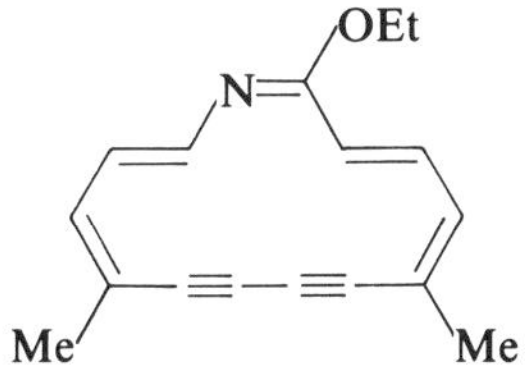

The aza[14]-, -[18]- and -[22]annulenes are diatropic; the [16]- and [20]-ring analogues are paratropic. In each case the ring current appears to be less, the greater the size of the ring. As in the case of annulenes, as the ring sizes increase there appears to be an alternation from $4n$ to $(4n + 2)$ rings in the wavelengths of the main absorption peaks in the electronic spectra. Tetra-aza[32]- and -[40]dehydroannulenes have also been made. Both are atropic, presumably because of the size of the rings.

A number of bridged aza-annulenes are known. Examples of methano-bridged, oxido-bridged and ethano-bridged aza-annulenes are as follows:

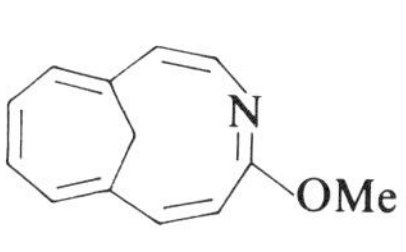

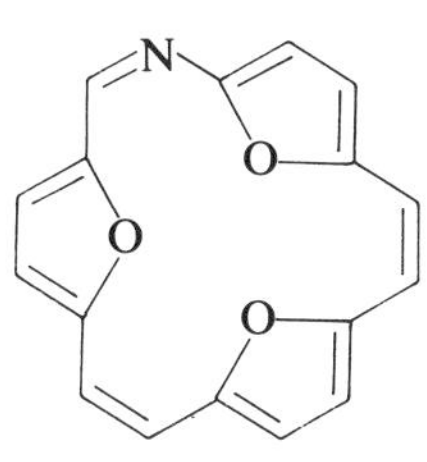

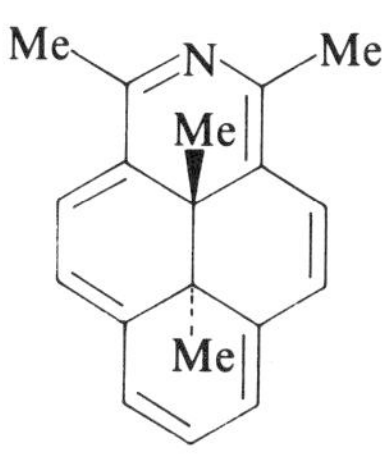

The latter two examples, as [18]- and [14]annulenes, respectively, are diatropic. The trioxido-aza-annulene is protonated in trifluoro-acetic acid and the n.m.r. signals for the protons are shifted even further downfield to δ 9.41 − 10.15.

The [1]H-n.m.r. spectrum of the methano-aza[12]annulene shows that it has a non-planar structure and localized single and double bonds. This compound is surprisingly stable and can be kept in air. This stability has been ascribed partly to relief of strain in the molecule by its taking up a non-planar form and partly to homoaromatic stabilization in the cycloheptatriene portion of the molecule.

An aza-annulene system of both importance and complexity has as its parent the compound porphin (or porphyrin):

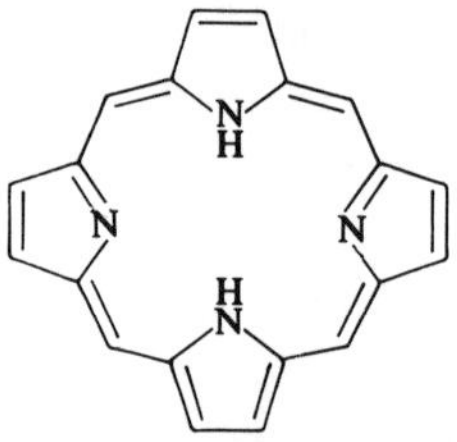

Derivatives of porphin are extremely important in nature, for they include chlorophyll, the green pigment of plants which is involved in photosynthesis, and haem, which forms part of haemoglobin present in the red blood corpuscles and acting as the oxygen carrier in the body. The important pigments known as the phthalocyanins are dibenzoporphins. All these compounds are metal complexes, with a metal atom in the centre of the ring (magnesium in the case of chlorophyll and iron in haem). N.m.r. spectroscopy shows clearly that this ring system is diatropic. This is most commonly represented as a diaza[18]annulene bridged by two ethylene groups and two amino groups, as in (A) below. In the following formulae the bold lines denote the possible cyclic conjugated systems:

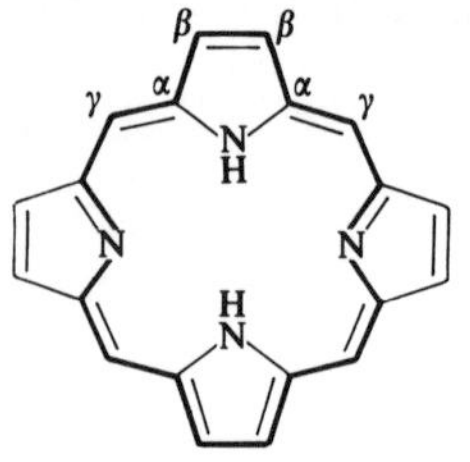

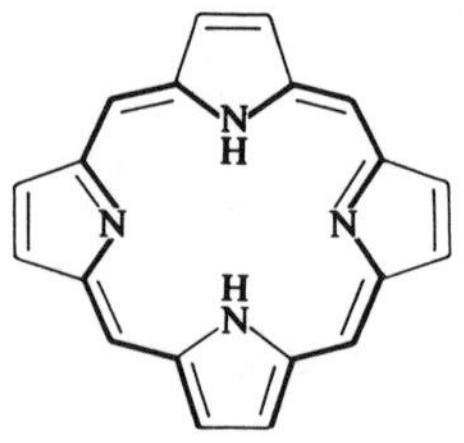

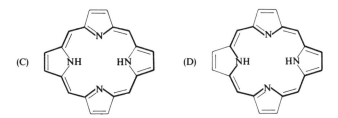

There is, of course, another Kekulé form of (A) with all the formal single and double bonds shifted one place around the ring. An alternative conjugated 18-π-electron ring is possible, as shown in (B). This is a bridged triaza[17]annulene, with a nitrogen atom, contributing two electrons, in place of a double bond. In addition, it is evident that prototropic shifts of the inner hydrogen atoms between the nitrogen atoms take place, so that (C) and (D) also contribute to the overall structure. In other words, a number of circuits of 18 π-electrons may contribute to the overall electronic structure of the molecule, and the resultant structure probably represents an average (not necessarily equally weighted) of such forms. This would lead to the expectation that bonds $\beta\beta$ might be less delocalized and have more double-bond character, and, concomitant with this, that bonds $\alpha\beta$ should have more single-bond character. Bonds $\alpha\gamma$ should, however, resemble normal aromatic delocalized bonds. This is in accord with X-ray studies, which show bond lengths of 1.373–1.398 Å ($\alpha\gamma$), 1.425–1.465 Å ($\alpha\beta$) and 1.344–1.371 Å ($\beta\beta$). It should be emphasized that the electronic pattern may change considerably in the metal complexes, since bonding between the porphin ring and the metal atom is also involved.

This brief allusion to the problems in trying to define the electronic structure of porphin may also serve as a general caution that situations may be complex, and that all possible factors must be taken into consideration. Even after that, there may be further unrecognized factors which also play a part.

Oxygen analogues of the porphins have been prepared. These are doubly charged cations:

$2X^-$

The salts are violet and are virtually insoluble in common organic solvents and in water; in the latter they undergo decomposition in a surface reaction. As with the porphins, a number of 18-π-electron circuits are possible. The $\alpha\gamma$ and $\beta\beta$ bond lengths resemble those in porphin but the $\alpha\beta$ bonds are noticeably shorter ($\sim$ 1.405 Å)

Isomers of porphins and of its oxygen analogue, with different bridging between the pyrrole or furan rings, have been prepared, e.g.

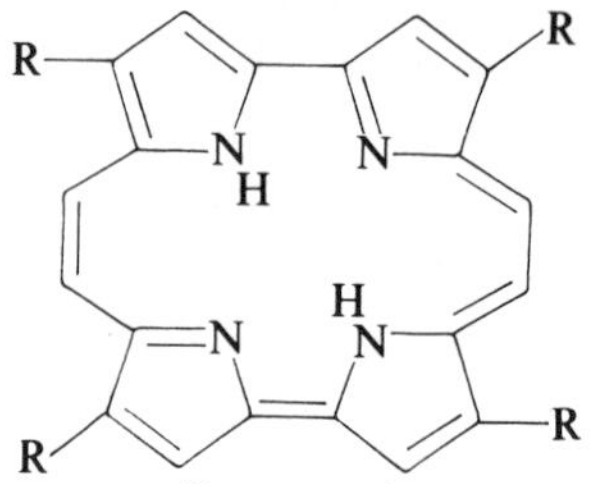

R = propyl

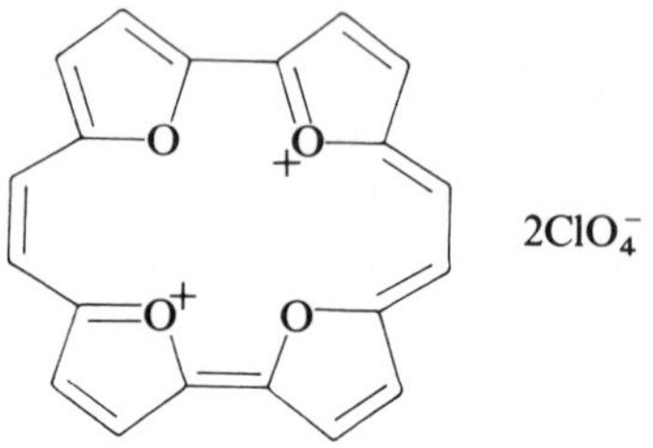

Each of these compounds has 18-π-electron circuits and is diatropic.

An analogue of porphin which can be regarded as a bridged diaza[34]annulene has been prepared:

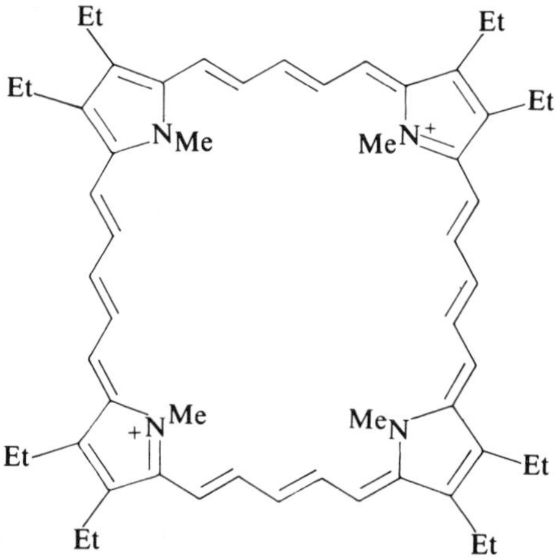

It is dark blue, very stable and supports a very strong diamagnetic ring current.

The diatropicity of porphin vinylogues appears to be greater than that of the corresponding annulenes, presumably because of a flattening effect due to the pyrrole rings, which hinders inversions in the annulene rings.

Möbius Annulenes

In benzene and other $[4n + 2]$annulenes there is a continuous array of p orbitals:

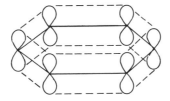

It has been suggested that if an annulene had instead a 180° twist in it, as in a Möbius strip, for such systems Hückel's rule would be reversed, i.e. $[4n]$ Möbius annulenes would be stabilized and aromatic whereas $[4n + 2]$ Möbius annulenes should be anti-aromatic. The shape of such a system would be as follows:

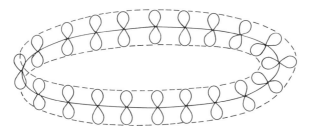

No such system has yet been synthesized. However, the concept is realized in the 'aromatic transition state' concept for pericyclic reactions whereby, for example, electrocyclic reactions involve Hückel-like transition states for reactions involving $(4n + 2)$-π-electron systems and Möbius-like transition states for $4n$-π-electron systems. (see also Chapter 14)

It might prove less difficult to construct a bridged Möbius annulene than a monocyclic example, but a stabilized $[4n]$ Möbius annulene may confidently be expected so long as strain does not override the electronic stabilization.

Further Reading

H. P. Figeys, 'Electronic structure and spectral properties of annulenes and related compounds', in *Topics in Carbocyclic Chemistry*, Vol. 1, Logos Press, London, 1969, pp. 269–359.

Dehydroannules: M. Nakagawa in *Topics in Nonbenzenoid Aromatic Chemistry*, Vol. 1, John Wiley, New York, 1973, pp. 191ff.

Annulene cations and anions: K. Müllen, *Chem. Rev.*, **84**, 603 (1984); *Angew. Chem.*, **99**, 192 (1987); *Angew. Chem. Int. Edn Engl.*, **26**, 204 (1987).

Cyclazines: D. Leaver, *Pure Appl. Chem.*, **58**, No. 1, 143 (1986); W. Flitsch and U. Krämer, *Adv. Het. Chem.*, **22**, 322 (1978).

Dimethano- and ethano-[14]annulenes: E. Vogel, *Pure Appl. Chem.*, **28**, 355 (1971).

See also Chapter 7 of the author's 1984 book (for details see Preface).

7
Derivatives of Cyclopentadiene

Cyclopentadiene is a typical reactive conjugated diene. It readily undergoes cyclo-addition reactions, including dimerization:

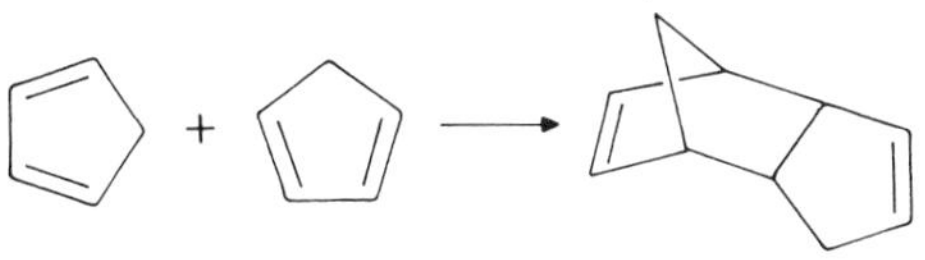

Its methylene group is weakly acidic and, as long ago as 1901, Thiele showed that in solution in dry benzene under dry nitrogen it reacted with potassium to form a yellow precipitate of potassium cyclopentadienide:

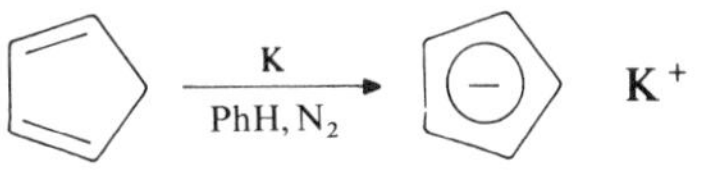

Since then many other salts having alkali metal, alkaline earth metal and rare earth metal cations have been prepared. These salts are extremely sensitive to air and moisture, sufficiently so that they may inflame in air. In an inert atmosphere they are, however, surprisingly stable; the sodium salt remained unchanged when heated at $300\,°C$ in nitrogen for a long period. The ionic nature of these salts has been confirmed. Thallium(I) cyclopentadienide is relatively stable in air, which makes it a useful reagent, but even this salt is oxidized in air sufficiently vigorously that it chars paper on which it is placed.

Cyclopentadiene is also sufficiently acidic to react with simple alkyl Grignard reagents, displacing the alkyl group and forming a cyclopentadienylmagnesium derivative:

$$C_5H_6 + C_2H_5MgBr \rightarrow C_5H_5MgBr + C_2H_6$$

Structure and Spectra of Cyclopentadienide Salts

As discussed in Chapter 2, the stability of the cyclopentadienide anion is associated with its having a delocalized sextet of π-electrons, four derived from the two double bonds of cyclopentadiene and two from the C—H bond which is cleaved when the ion is formed. Prior to the enunciation of Hückel's rule this connection had been recognized in 1928: 'In regard to its ability to provide the electrons for the stable sextet—cyclopentadiene can do so only by appropriation of the electrons of one of its hydrogen atoms, it is this circumstance which gives to the hydrocarbon and its derivatives properties analogous to those of an acid, and confers stability on the corresponding anion' (F. R. Goss and C. K. Ingold, *J. Chem. Soc.*, 1268 (1928)).

The pK_a of cyclopentadiene in water, in which it is slightly soluble, is 16; its acidity is thus comparable to that of water and alcohols rather than of most hydrocarbons.

The symmetry of the cyclopentadienide anion is shown by its n.m.r. spectra, which have single peaks at about δ 5.5 (^{1}H) and 103.8 (^{13}C). Both peaks appear upfield from the corresponding peaks provided by benzene; this can be accounted for by the charge on the anion. Because benzene and the cyclopentadienide anion each have six π-electrons and the radii of the two rings are similar, the resultant deshielding due to the diamagnetic ring current is almost the same in each case, and the different chemical shifts in a common solvent are due almost entirely to the different electron densities on the rings in the two species.

The ^{7}Li-n.m.r. spectrum of lithium cyclopentadienide in tetrahydrofuran suggests that the lithium ion is located over the π-cloud of the anion, forming a tight or contact ion pair, but in the more polar solvent hexamethylphosphoric triamide (HMPA) the ions are solvated entities.

Chemical Reactions

Not surprisingly, in solution the cyclopentadienide ion reacts readily with electrophiles, although it has been described as a poor nucleophile in the gas phase. It reacts with acid chlorides without a catalyst being needed to give diacyl or diaroyl derivatives, and with carbon dioxide to give a dicarboxylic acid which dimerizes. Alkyl halides also react to give dimeric alkyl derivatives, e.g.

Substituted Cyclopentadienide Salts

The presence of electron-withdrawing substituents on the cyclopentadienide ring greatly lowers the reactivity of the salt and these salts can be kept in air. This is because the negative charge is delocalized into the substituent groups as well as in the ring. Such salts include acyl, aroyl, methoxycarbonyl, cyano and methylsulphonyl derivatives.

The presence of electron-withdrawing substituents greatly increases the acidity of the cyclopentadiene ring. For example, the pK_a values in aqueous solution of methoxycarbonyl- and nitro-cyclopentadiene are, respectively, 10.35 and 3.25. Pentasubstituted cyclopentadienes of this sort are extremely strong acids. Thus in solution in acetonitrile the pentacyanocyclopentadienide anion is not protonated by perchloric acid.

Polysubstituted cyclopentadienide salts may none the less undergo substitution reactions with electrophiles. The tetracyano derivative can be halogenated, nitrated and acetylated, but in this case acylation requires the presence of an acid catalyst.

Indene and fluorene (benzo- and dibenzocyclopentadiene) also have acidic methylene groups and provide indenide and fluorenide salts. However, annelation of the benzene rings reduces the acidity of the cyclopentadiene ring, implying reduced stabilization of the cyclopentadienide ion relative to the hydrocarbon in these cases. In Chapter 11 it will be seen that benzo-annelation commonly lowers the stabilization of cyclic polyalkenes and related salts.

Cyclopentadienylides

The negatively charged cyclopentadienide ring may occur not only as a discrete anion but also as the negatively charged part of a dipolar molecule, as in

Several types of compounds involve contributions of differing extent from this type of dipolar structure. In *cyclopentadienylides* the negatively charged cyclopentadienide ring is attached directly to a positively charged heteroatom, as in trimethylammonium cyclopentadienylide:

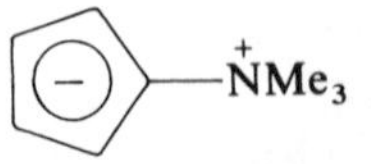

If the hetero-atom is a member of the second or lower rows of the periodic table, as, for example, in triphenylphosphonium cyclopentadienylide (A), then an alternative structure (B) may be drawn, since the hetero-atom may expand its valence shell, utilizing unfilled *d*-orbitals to achieve this:

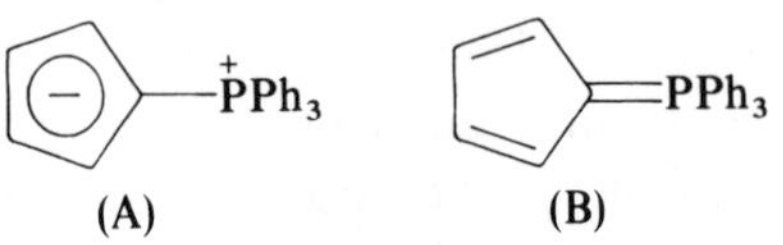

(A)   (B)

Such molecules exist as hybrids of these two possible structures.

The first cyclopentadienylide to be prepared was pyridinium cyclopentadienylide, made from cyclopentadiene as follows, in 1955:

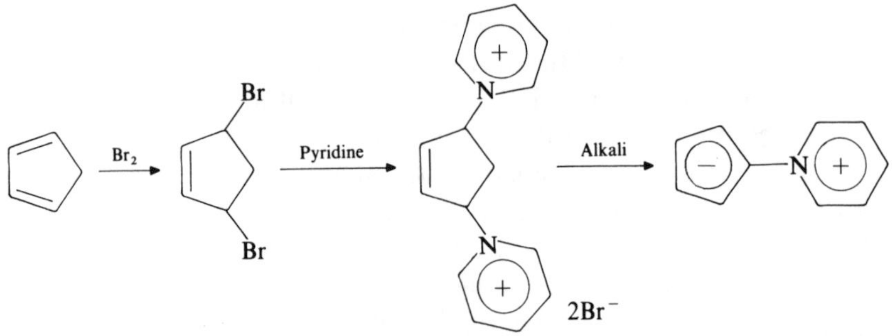

A number of different cyclopentadienylides, with a variety of hetero-atoms, has since been prepared.

Pyridinium cyclopentadienylide has a very large dipole moment (13.5D); that of triphenylphosphonium cyclopentadienylide is much smaller (7.0D), due to the contribution from the alternative structure (B). Spectroscopy and calculations suggest that the dipolar form (A) makes ∼ 80% contribution to the structure of the phosphonium ylide.

Chemical Reactions of Cyclopentadienylides

Two types of chemical reactivity of cyclopentadienylides are of particular interest: their ability to undergo electrophilic substitution reactions in the five-membered ring and their participation as ylides in the Wittig reaction.

Electrophilic substitution takes place preferentially at the 2- and

5-positions, but with reactive electrophiles all sites may be attacked. The reactions mechanistically resemble those of benzene:

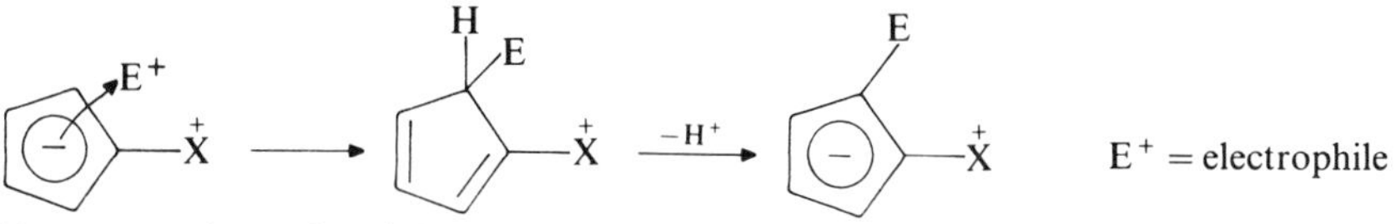

E^+ = electrophile

Regeneration of a delocalized six-π-electron system in the ring results, as in the case of benzene, thus bringing about overall substitution reactions. Some examples are as follows:

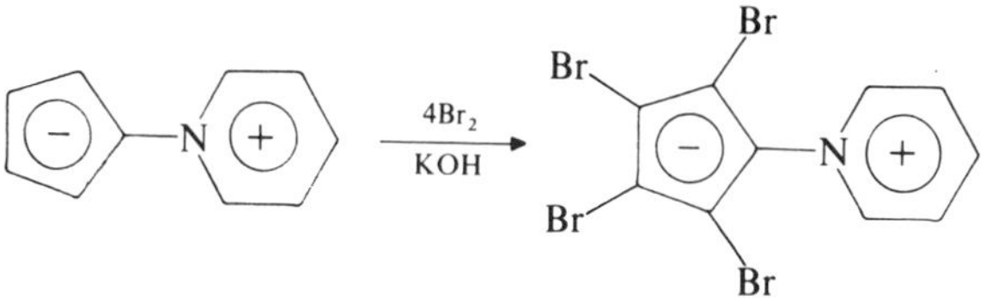

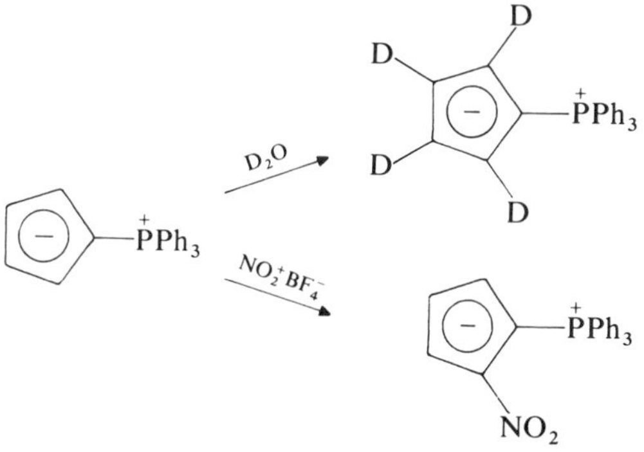

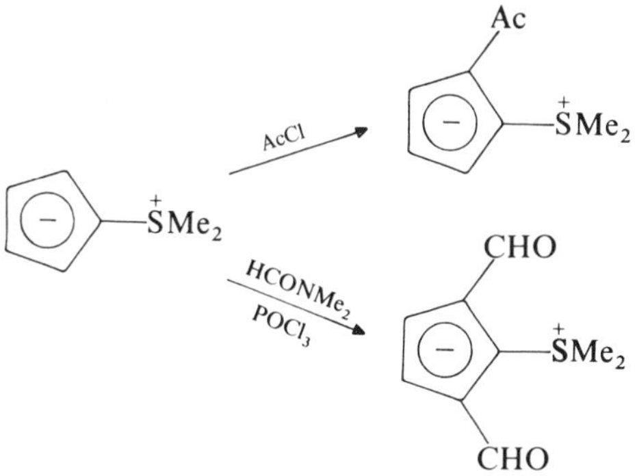

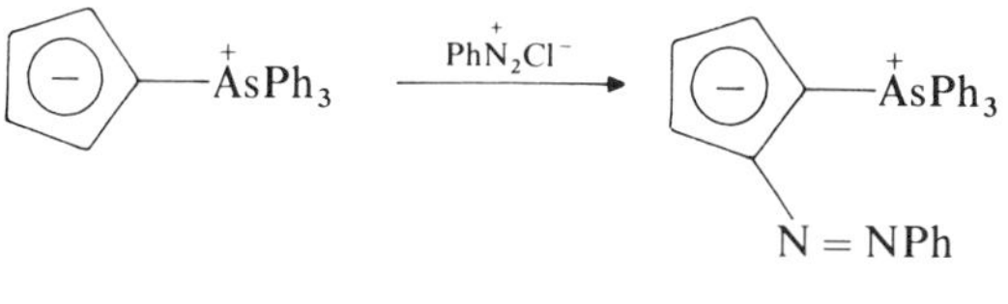

Electron-withdrawing substituents in the cyclopentadienide ring have the same effect as in a benzene ring and lower its reactivity towards electrophiles. In addition, the more readily the exocyclic heteronium group participates in double-bond formation with the ring, thus lowering the polarity of the ylide bond, the lower is the reactivity of the cyclopentadienide ring towards electrophiles.

This difference in reactivity of different cyclopentadienylides makes itself evident in their shelf-lifetime. Thus the highly polar pyridinium cyclopentadienylide is destroyed fairly quickly by contact with air (although it is stable when kept under nitrogen), while its less polar triphenylphosphonium analogue can be kept for long periods in air. In general, from the second row of the periodic table downwards the reactivity of related ylides increases as does their susceptibility to the atmosphere.

In parallel with their reactivity towards electrophiles is their basicity, which depends upon the relative stability of the ylides and the cations formed when they are protonated:

The stability of the ylides, which is related to the extent of double-bond formation with the exocyclic hetero-atom, largely determines this equilibrium, and the basicity increases down the periodic table. Protonation may take place at any of the sites in the ring and is affected by the substitution pattern in the ring. Thus dimethylsulphonium cyclopentadienylide in trifluoroacetic acid is protonated at the 1-, 2- or 3-positions in the ratios 11:56:33.

There have been no reports of Wittig reactions involving unsubstituted cyclopentadienylides. However, Wittig reactions of a variety of phenyl-substituted cyclopentadienylides have been recorded, e.g.

Reactivity in the Wittig reaction also depends on the polarity of the ylide bond; for example, whereas the above reaction provides a 95% yield, the triphenylphosphonium analogue does not react with *p*-nitrobenzaldehyde. Electron-withdrawing substituents in the five-membered ring, which delocalize the negative charge out of the ring, also inhibit the reactions of cyclopentadienylides with aldehydes.

Diazocyclopentadienes

Diazocyclopentadienes are closely related to cyclopentadienylides. Their overall structure is a hybrid of two structures, one having a cyclopentadienide ring and the other with the negative charge on the terminal nitrogen atom:

Diazocyclopentadiene is formed by reaction of lithium cyclopentadienide with toluene-*p*-sulphonazide:

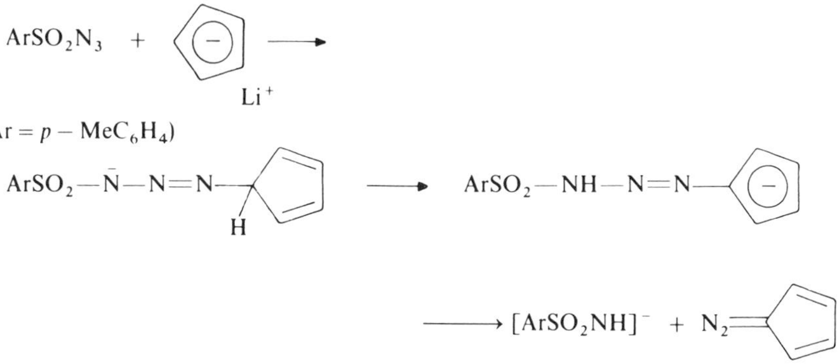

$(Ar = p - MeC_6H_4)$

Other substituted diazocyclopentadienes have been prepared in the same way and also by oxidation of hydrazones, e.g.

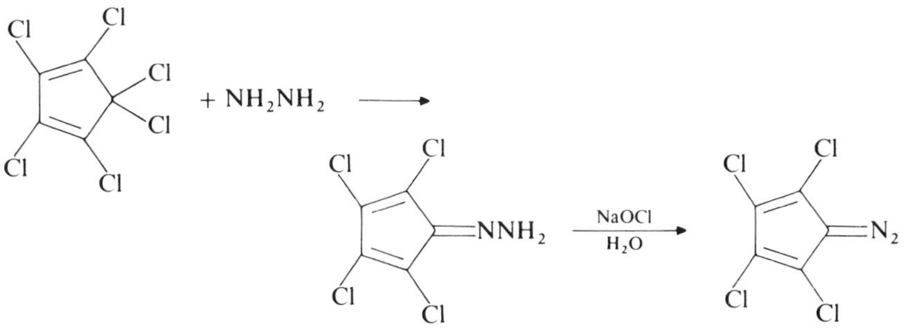

and, in the case of some highly substituted derivatives, from the corresponding amine, e.g.

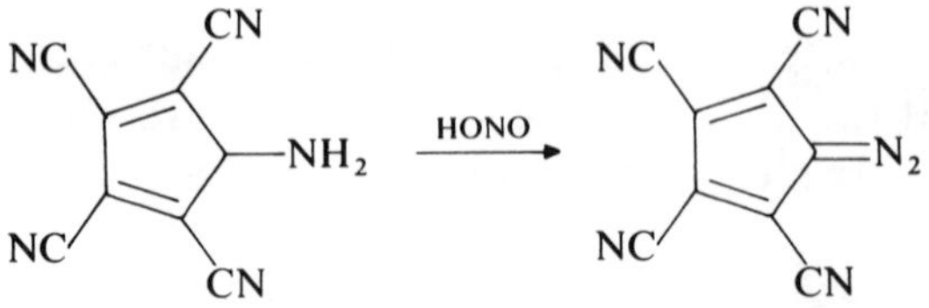

Spectroscopic data are inconclusive as to which of the two possible contributing forms is the more important and what is the precise structure of diazocyclopentadiene; it is best to consider it as a hybrid of the ylide and diene structures.

Since its cyclopentadiene ring is electron-rich, diazocyclopentadiene would be expected to react with electrophiles, and, since the five-membered ring is stabilized by the contribution of a delocalized six-π-electron system, substitution reactions are again the result. These reactions take place preferentially, though not exclusively, at the 2,5-positions, as in the case of cyclopentadienylides. Thus diazo-coupling takes place at position 2, but nitration (by benzoyl nitrate) at positions 2 or 3, with the former preferred. Bromination leads to a tetrabromo derivative:

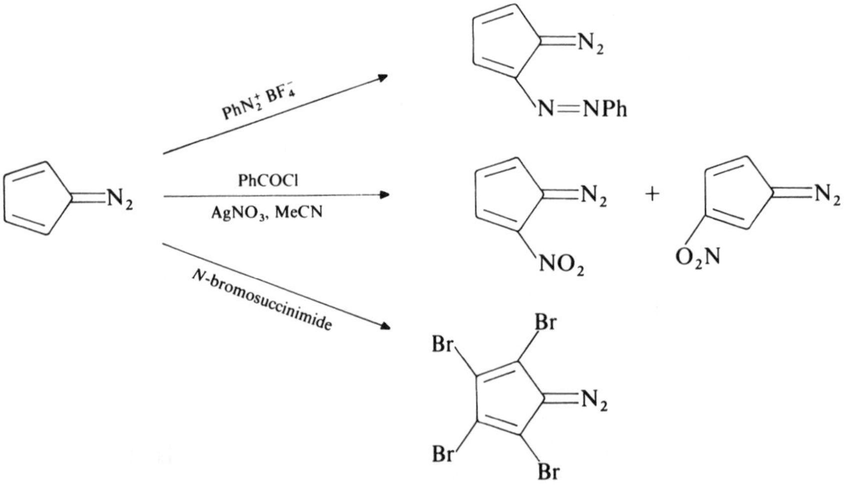

Like other diazo compounds, diazocyclopentadienes can lose nitrogen to generate a carbene when heated or irradiated by light. These carbenes undergo typical carbene reactions such as addition to alkenes, generating spiro[4.2]dienes, e.g.

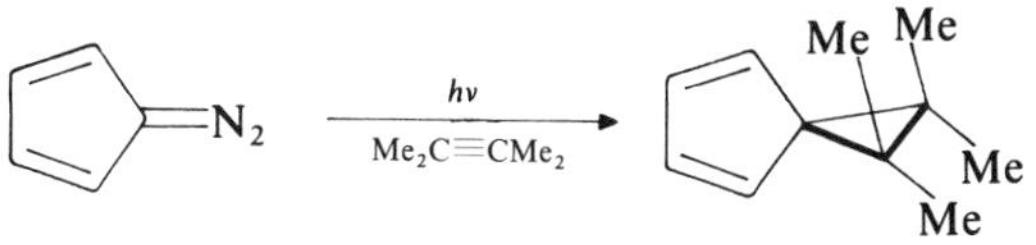

The carbenes can also be trapped by derivatives of Groups V and VI (or 15 and 16) elements to form cyclopentadienylides with a variety of exocyclic hetero-atoms, and this is a very useful method for the preparation of these ylides. The following cyclopentadienylides have been prepared in this way from diazotetraphenylcyclopentadiene:

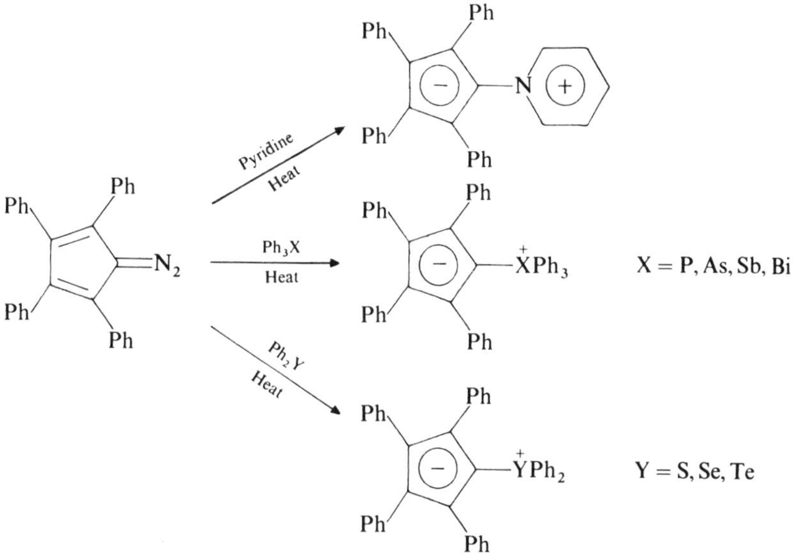

$X = P, As, Sb, Bi$

$Y = S, Se, Te$

Fulvenes

Cyclopentadiene and substituted cyclopentadienes react with aldehydes or ketones in the presence of a base to form strongly coloured, usually orange or red, methylenecyclopentadienes called *fulvenes*:

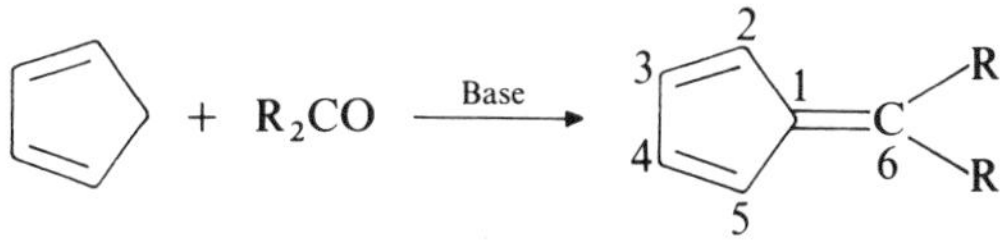

The parent compound, fulvene $(R = H)$, is highly reactive and polymerizes and undergoes autoxidation very readily. Substitution at the 6-position, and especially by aryl groups, lowers the reactivity and increases the longevity of fulvenes.

Although there is no hetero-atom in simple fulvenes, it is still possible that a dipolar structure

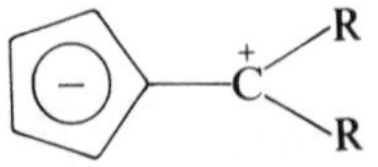

might make some contribution to the overall structure, since it would provide a ring with a delocalized sextet of π-electrons. However, all the structural and spectroscopic evidence suggests that simple fulvenes are best represented by formulae with defined single and double bonds, and that a dipolar structure makes, at most, no more than 10% contribution to the overall structure. The dipole moment of fulvene is only 0.42D. The dipolar contribution is thought to be larger in the excited state, and to be responsible for a decrease in the energy difference between the ground and excited states, thereby causing fulvenes to absorb light at longer wavelengths than do isomeric benzenoid compounds, and to be coloured.

Fulvenes behave chemically as reactive alkenes. Simple fulvenes readily undergo oxidation and polymerization and participate in cyclo-addition reactions both as dienes and as dienophiles:

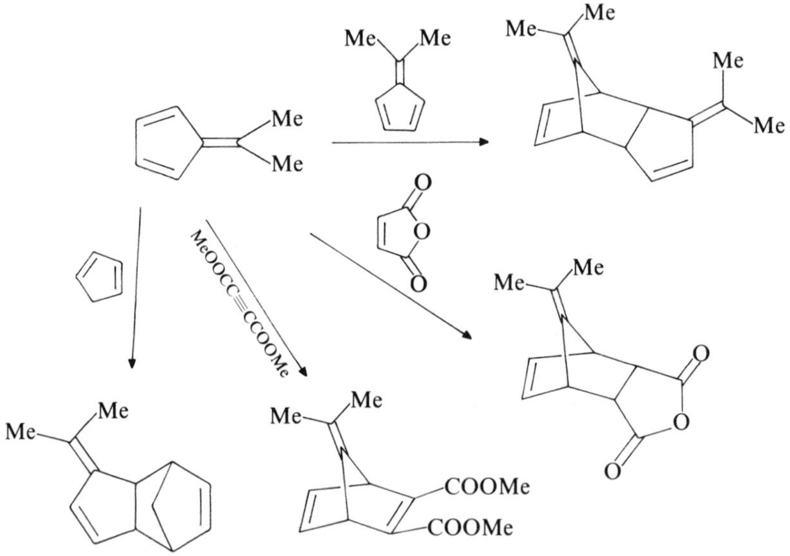

They undergo addition reactions with halogens. With sodamide cyclopentadienide salts are formed:

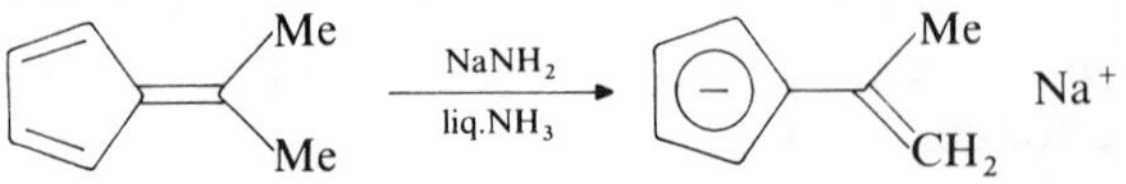

The presence of a hetero-atom substituent at position 6 may increase the polarity of fulvenes and also their stability. This increase in polarity is demonstrated by the dipole moment, 4.5D, of 6-dimethylaminofulvene. The greater polarity is due to electronic interaction between the nitrogen atom and the five-membered ring, leading to greater development of cyclopentadienide character in the ring, e.g.

Another consequence of this interaction is a partial double-bond character in the C(6)—N bond, which results in rotation about this bond being restricted. As a result of this, n.m.r. spectra of the above 6-aminofulvene recorded at low temperatures provide two distinct signals for the two *N*-methyl groups, although these signals coalesce at room temperature.

These 6-aminofulvenes do not undergo cyclo-addition reactions with dienophiles. They are, however, activated towards attack by electrophiles such as bromine. This happens because they are enamines. The amino-group itself is readily replaced by nucleophilic attack:

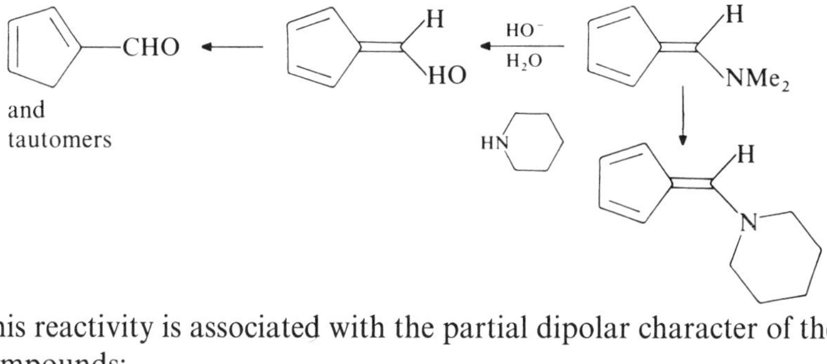

This reactivity is associated with the partial dipolar character of these compounds:

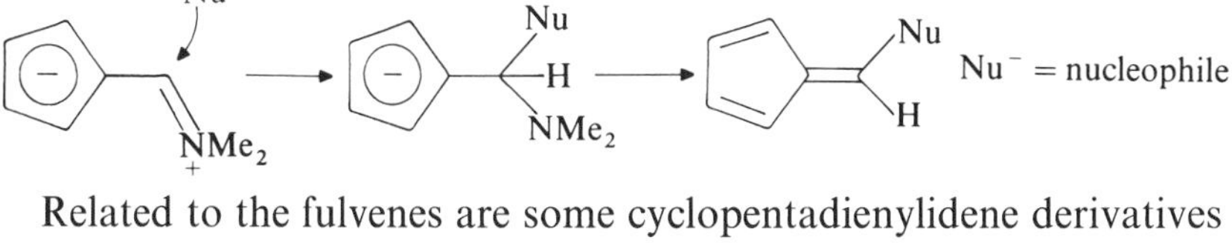

Related to the fulvenes are some cyclopentadienylidene derivatives of heterocycles:

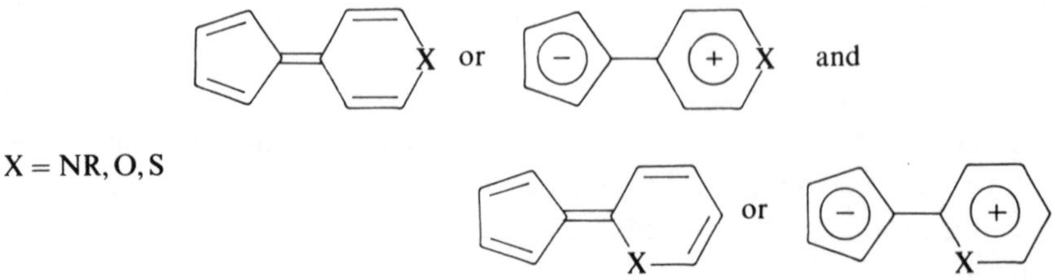

X = NR, O, S

These compounds might be represented either as fulvenes or as dipolar compounds resembling cyclopentadienylides. Spectroscopic and crystallographic evidence suggests that they exist as hybrids of these two extreme structures, but with a preponderance of fulvene-like character. The dipolarity is greatest when X = NR.

A number of electrophilic substitution reactions have been shown to take place in the five-membered rings of these compounds, e.g.

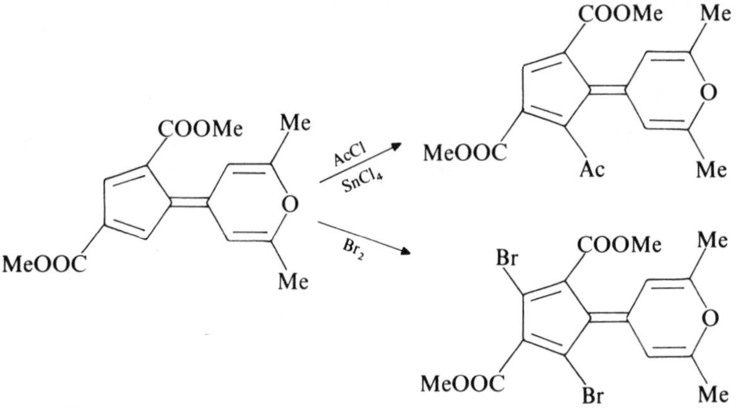

Such reactions are not, however, indicative of cyclopentadienide character in the five-membered ring but rather may reflect the conjugative interaction of the hetero-atom with the reacting site, leading to a stable heteronium ion intermediate:

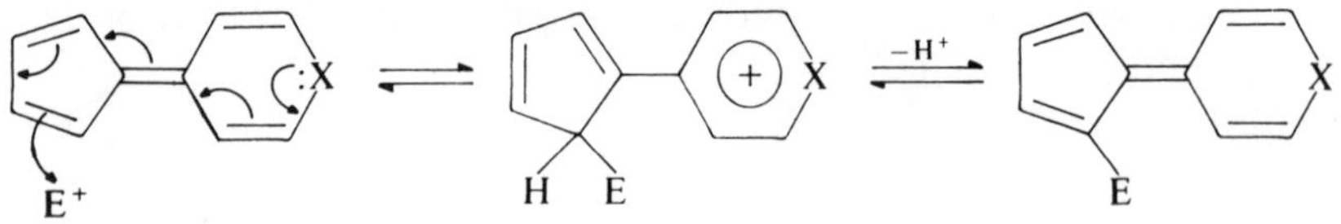

Ferrocene

This book in general does not discuss organometallic compounds, but to consider cyclopentadiene derivatives without mention of ferrocene, $(C_5H_5)_2Fe$, is akin to *Hamlet* without the Prince, so important is the role it has played in this field of chemistry.

The iron atom of ferrocene is sandwiched between two C_5H_5 rings:

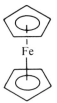

In its crystalline state the cyclopentadienyl rings of ferrocene are staggered, as shown in the above formula. The barrier to rotation of the rings is, however, low, and in solution they rotate freely.

The discovery of ferrocene is a superb example of the role of serendipity.* In the early 1950s one group of workers was trying to make fulvalene (see Chapter 10) by allowing cyclopentadienyl magnesium iodide to react with iron (III) chloride. A separate group was studying the reactions of alkenes with nitrogen over catalysts and investigated the reaction of cyclopentadiene with nitrogen in the presence of iron at 300 °C. Neither group achieved its objective. Both prepared a novel orange crystalline compound, $C_{10}H_{10}Fe$. This was remarkably stable, being insoluble in, and unaffected by, water, solutions of sodium hydroxide or concentrated hydrochloric acid, even at their boiling points. It sublimed unchanged at 100 °C, and was stable up to 500 °C.

Shortly after its initial preparation it was assigned its now-accepted sandwich structure. Originally it was regarded roughly as a ferrous ion sandwiched between two negatively charged cyclopentadienide rings, but now it is realized that the situation is much more complicated than that. An alternative extreme description would have two neutral cyclopentadienyl rings bonded to a neutral iron atom. The detailed picture of the bonding is complex; the latter description is nearer to the overall one. In particular, it should be noted that the overall count of bonding electrons (in the extreme structures, either 12 from $2(C_5H_5)^-$ plus 6 from Fe(II), or 10 from $2(C_2H_5)$ plus 8 from Fe°) is 18, giving the krypton inert gas structure.

Ferrocene is now a compound of wide interest and importance, and was the first known representative of a large class of related compounds in which one or more cyclopentadienyl rings are π-bonded

*'Serendipity' is used intentionally rather than 'luck'.

to a central transition metal atom. As well as sandwich compounds such as ruthenocene, $(C_5H_5)_2Ru$ and the cobalticinium cation $(C_2H_5)_2Co^+$, other related compounds have a cyclopentadienyl ring on one side of the metal and other ligands forming the other side of the sandwich, e.g.

Ferrocene is now commonly made by the reaction of cyclopentadiene with iron(II) chloride in the presence of a base (see *Org. Prepns*, **36**, 31, 34 (1956)).

Chemistry of Ferrocene

A large amount of work has been carried out on the chemistry of ferrocene but only a few salient points relevant to its relationship with cyclopentadienide systems will be mentioned here.

Ferrocene is readily acylated in Friedel–Crafts reactions; mild conditions suffice. Introduction of an acyl group deactivates the whole molecule to electrophilic attack, especially the ring to which it is attached, so that substitution of a second acyl group occurs in the hitherto unsubstituted ring:

Ferrocene is sulphonated using acetic anhydride as solvent.

Attempts to brominate or nitrate ferrocene directly are foiled since, instead of electrophilic substitution in the rings, oxidation of the molecule to a blue *ferricinium* cation $[(C_5H_5)_2Fe]^+$ takes place. This cation resists electrophilic attack. Unlike ferrocene, the ferricinium cation is soluble in water. It is paramagnetic, in keeping with its having an unpaired electron, and is readily reduced back to ferrocene. Ferrocene itself is very resistant to catalytic reduction, but is reduced by lithium in ethylamine to iron and cyclopentadiene.

Cyclopentadienium Salts

Removal of a hydride ion from cyclopentadiene would leave a cyclopentadienium cation:

$$\text{(cyclopentadiene)} \xrightarrow{-\,H^+} \text{(cyclopentadienium cation, +)}$$

This cation has only four π-electrons, and thus from Hückel's rule should not have any special stabilization and may actually be destabilized as a $4n$-π-electron system.

Circumstantial evidence for its instability compared to the cyclopentadienide anion derives from how much more elusive it proved to be to obtain. It was eventually generated in the 1970s by reaction of 5-bromocyclopentadiene with antimony pentafluoride at low temperature. Its e.s.r. spectrum indicates that it has a triplet ground state.

Heterocyclic Analogues of the Cyclopentadienide Anion

Examples of heterocyclic analogues of cyclopentadiene are the well-known compounds pyrrole, thiophen and furan:

| Pyrrole | Thiophen | Furan |

All three of these compounds have, in their neutral form, possible sextets of electrons, made up of four electrons from the two double bonds and two electrons from the hetero-atoms.

The carbon–carbon bond lengths in each of these compounds differ from those in benzene or in pyridine in that they alternate in length:

1.417 / 1.382 (N–H)   1.423 / 1.370 (S)   1.431 / 1.361 (O)   (Å)

The 3,4-bonds obviously have more single-bond character and the 2,3- and 4,5-bonds more double-bond character. Delocalization of the six electrons depends on the extent to which the two electrons contributed by the hetero-atoms participate in this delocalization, and on the extent to which they remain associated with the hetero-atoms.

Consideration of the bond lengths would indicate that delocalization diminishes in the order pyrrole > thiophen > furan.

All three compounds are electron-rich, having six π-electrons among five ring atoms. Consequently they are all readily attacked by electrophiles; they are indeed more reactive than benzene. An alternative way of considering this reactivity is to regard pyrrole, for example, as an enamine. Electrophilic attack might take place at either the 2- or 3-positions:

E^+ = electrophile

Of the alternative positively charged intermediate species, the cation derived from attack at the 2-position can have greater delocalization of the positive charge than is possible for the other cation:

The former cation is thus likely to be the more stable with a lower transition state leading to its formation. Substitution does indeed take place preferentially at the 2-position. The difference between reactivity at the 2- and 3-positions is most marked in furan and least so in pyrrole.

Pyrrole is by far the most reactive of the three heterocycles to electrophilic attack, reflecting the greater electron-donating character of nitrogen compared to oxygen or sulphur. Furan is more reactive than thiophen. All three heterocycles undergo substitution reactions, the regeneration of their ring structures testifying to the stability of these ring systems.

Pyrrole is an extremely weak base, pK_a -3.8, in contrast to pyridine, discussed in Chapter 1, with pK_a 5.17. The contrast between these two cyclic amines derives largely from the fact that pyridine has a lone pair of electrons, not forming part of its delocalized sextet of π-electrons, which is available for reaction with an acid, whereas the sextet in pyrrole includes the lone pair of electrons, and their utilization in protonation disrupts the sextet and thus costs energy. Reversible protonation occurs at any site in the ring. Protonation on nitrogen is kinetically favoured, while protonation at C(2) or C(5) is thermodynamically favoured. Protonation at either site involves the lone pair of electrons formally associated with the nitrogen atom:

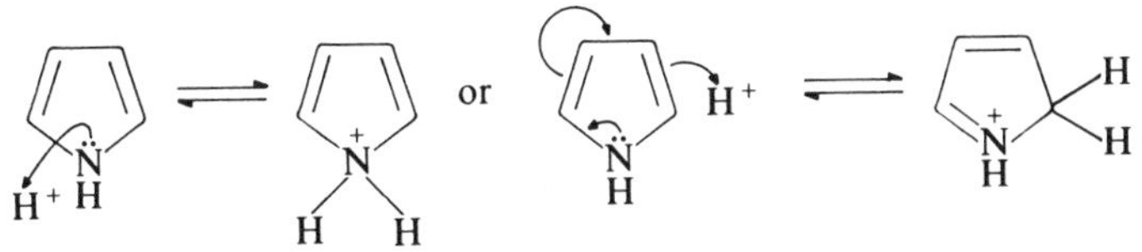

Another interesting example of the role played by the lone pair of electrons on the hetero-atom in these heterocycles is provided by the effect of oxidation of thiophen by peroxybenzoic acid, which gives rise to dimeric products. In formation of the dioxide, the sextet of π-electrons is lost and the resultant diene reacts very readily in cyclo-addition reactions providing dimeric products.

Thiophen resembles benzene in being fairly resistant to oxidation; it is unaffected by aqueous potassium permanganate. In contrast, pyrrole and furan are very readily oxidized.

Furan reacts readily with dienophiles in cyclo-addition reactions. Pyrroles do not do so and nor do thiophens except with extremely reactive dienophiles such as dicyanoacetylene.

Like cyclopentadiene, pyrrole can lose a proton to give an aza-analogue of the cyclopentadienide anion. These salts are usually prepared either by reaction with potassium or sodium amides in liquid ammonia or with butyl lithium in ether:

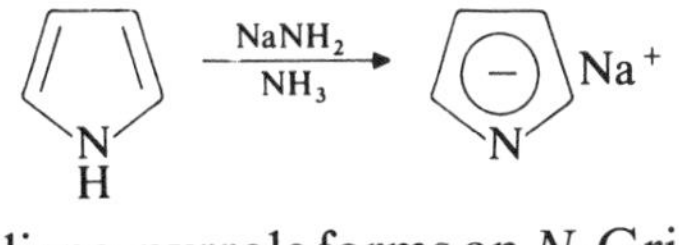

Also like cyclopentadiene, pyrrole forms an N-Grignard reagent. These metal salts react readily with alkyl and acyl halides.

More than one atom of the cyclopentadiene ring may be replaced by hetero-atoms, as in the following examples:

Pyrazole   Imidazole   Oxazole   Thiazole   Thiadiazole   Tetrazole

These *azoles* are related to pyrrole, furan and thiophen in the same way that pyridine is related to benzene and the diazines to pyridine, namely by replacement of a CH group by a N atom. This leads to compounds of lower reactivity. In many ways the chemistry of the azoles is an amalgam of that of the monoheterocyclopentadienes and that of pyridine. Imidazoles and thiazoles undergo electrophilic substitution only with difficulty. Pyrazole, imidazole and thiazole all resemble benzene or pyridine in their tendency 'to retain the type,' i.e. to undergo substitution rather than addition reactions; this is less the case with oxazoles. Introduction of an extra nitrogen atom increases the basicity of the heterocycles; the lone pair of electrons of this nitrogen atom are additional to the sextet of electrons in the ring and thus are available for salt formation.

Introduction of yet another nitrogen atom into the ring, as in thiadiazoles, oxadiazoles and triazoles, makes these rings even less reactive and they are very stable to oxidation, while electrophilic substitution reactions are virtually unknown. Tetrazole is similarly very resistant to oxidation. Compounds with numbers of nitrogen atoms linked together are commonly unstable, and the stability of tetrazole must be associated with its sextet of π-electrons.

Intriguing compounds reported in 1987 are the sodium and lithium pentaphosphacyclopentadienides:

$$\underset{P}{\overset{P-P}{P \ominus P}} \qquad Na^+ \ or \ Li^+$$

These compounds were made from white phosphorus and, respectively, sodium or lithium phosphide. They are extemely sensitive to oxidation but are stable for some days in dilute solutions at room temperature. This is the first reported homocyclic aromatic system wherein the ring atoms are not carbon.

Further Reading

Cyclopentadienides: H. J. Lindner, in *Methoden der Organischen Chemie (Houben–Weyl)*, 4th edition, Vol. 5/2c, Thieme, Stuttgart, 1985, pp. 45ff.

Cyclopentadienylides: G. Becker, in *Methoden der Organischen Chemie (Houben Weyl)*, 4th edition, Vol. 5/2c, Thieme, Stuttgart, 1985, pp. 494ff.

Fulvenes: K-P. Zeller, in *Methoden der Organischen Chemie (Houben–Weyl)*, 4th edition, Vol. 5/2c, Thieme, Stuttgart, 1985, pp. 504ff. For earlier work on fulvenes see P. Yates, in *Advances in Alicyclic Chemistry*, Vol. II, Academic Press, New York, 1968, pp. 59ff., and E. D. Bergmann, *Chem. Rev.*, **68**, 41 (1968).

For ferrocenes and other related metal derivatives of cyclopentadiene (etc.) see S. G. Davies, *Organotransition Metal Chemistry: Application to Organic Synthesis*, Pergamon Press, Oxford, 1982.

See also Chapter 2 of the author's 1984 book (for details see Preface).

8

Derivatives of Cycloheptatriene

As mentioned in Chapter 2, Hückel, in 1931, predicted that the cation formed by loss of a hydrogen atom together with its bonding electrons from cycloheptatriene should be a stable species:

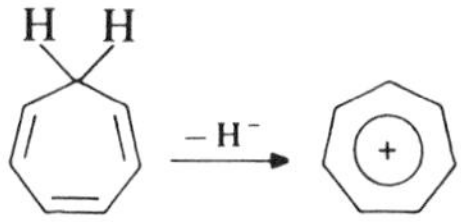

In fact this cation had been prepared, unintentionally, unwittingly and unwanted, forty years previously (G. Merling, *Ber.*, **24**, 3108 (1891)). Cycloheptatriene had been converted into its dibromo-adduct; when a distillation of the product was attempted the liquid was converted into a yellow solid, which was obviously not the product which was wanted. The work was repeated in 1954 and the yellow solid was identified as cycloheptatrienium bromide, or, to give it its usual name, *tropylium bromide*:

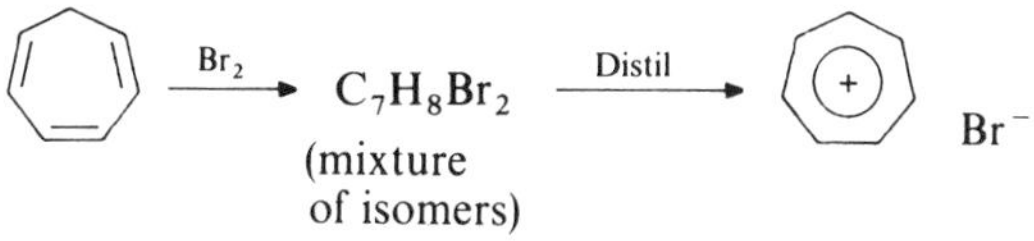

Not only Merling's work but also that of Hückel was originally largely overlooked by organic chemists, and interest in the chemistry of derivatives of cycloheptatriene was aroused in the 1940s by quite independent work carried out in England, Japan and Sweden on mould products (probably stimulated by the interest in penicillin) and on antibiotics obtained from the heartwood of *cupressaceae*.

In 1945 Dewar suggested that the mould product stipitatic acid was a dihydroxycycloheptatrienone carboxylic acid:

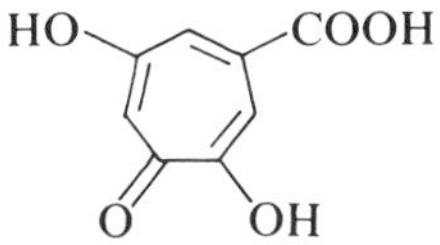

He also suggested that the cycloheptatrienolone ring might represent a new kind of aromatic system which would be stabilized by resonance and tautomerism:

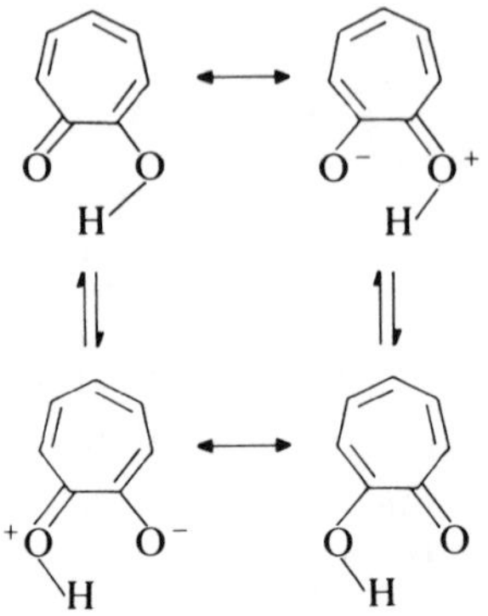

This ring was called *tropolone*, this name deriving from tropine, which is a breakdown product of the alkaloid atropine, obtained from deadly nightshade (*Atropa belladona*), and from which cycloheptatriene was first prepared, in 1891, by using a series of exhaustive methylation and other steps:

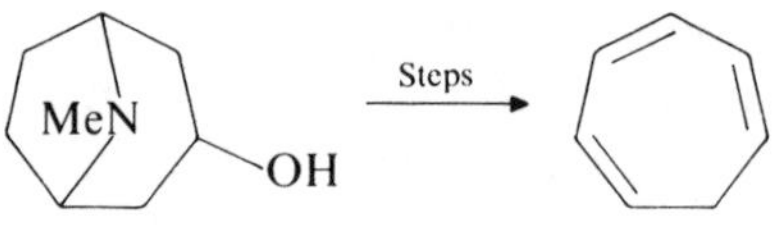

Tropine

About 1950 Hückel's rule began to have a general impact on organic chemistry, and this led to the consideration of possible dipolar structures for tropolones:

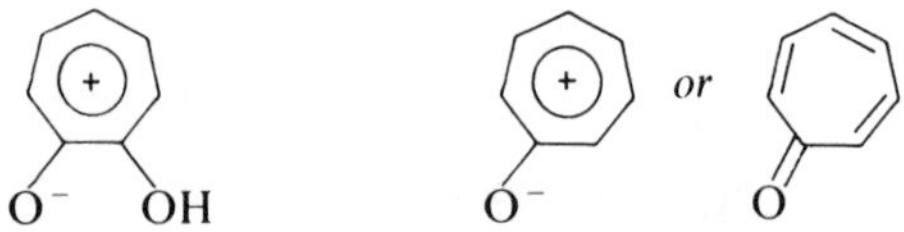

Tropone

This also led to the idea that the parent compound of the tropolones was tropone, rather than tropolone. It was thought that in these dipolar structures stabilization might be achieved by polarization of the carbonyl group, which would leave six electrons and a positive charge that could be delocalized over the seven-membered ring (as in a tropylium ion) and a negative charge on the oxygen atom. In other words, this involved a situation similar to that in cyclopentadienylides, but with the charges reversed, the positive charge being on the carbocyclic ring and the negative charge on an exocyclic group.

It was soon realized that the properties of tropone and tropolone were best interpreted without recourse to these dipolar structures. However, the initial association of 'aromaticity' with these compounds stimulated a very large amount of practical work and thought, not only in this series of compounds but also beyond.

Preparation of Tropylium Salts

The commonest methods presently used for the preparation of tropylium salts involve direct removal of a hydride ion from cycloheptatriene. Thus *Organic Syntheses, Collective Vol. 5* (1973), p. 1138 describes a method using phosphorus pentachloride:

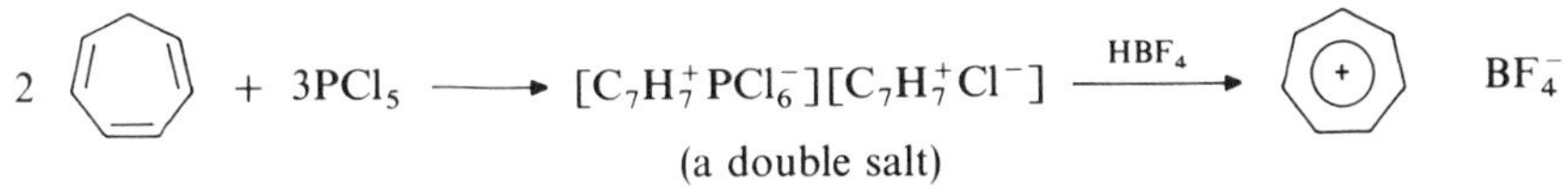

$$2 \; \text{(cycloheptatriene)} + 3PCl_5 \longrightarrow [C_7H_7^+ PCl_6^-][C_7H_7^+ Cl^-] \xrightarrow{\text{HBF}_4} \text{(tropylium}^+\text{)} \; BF_4^-$$

(a double salt)

Other reagents which have been used for this purpose include various triphenylmethyl salts and a mixture of ammonium nitrate and trifluoroacetic anhydride.

Tropyl ethers (alkoxycycloheptatrienes) are readily cleaved by acids to give tropylium salts in excellent yield:

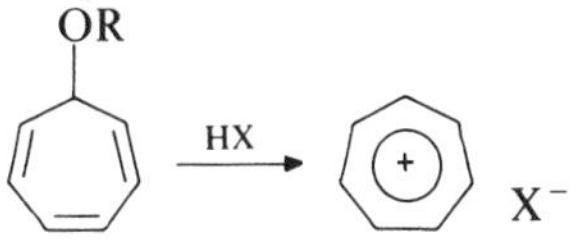

An interesting salt which has been prepared recently is a tropylium cyclopentadienide:

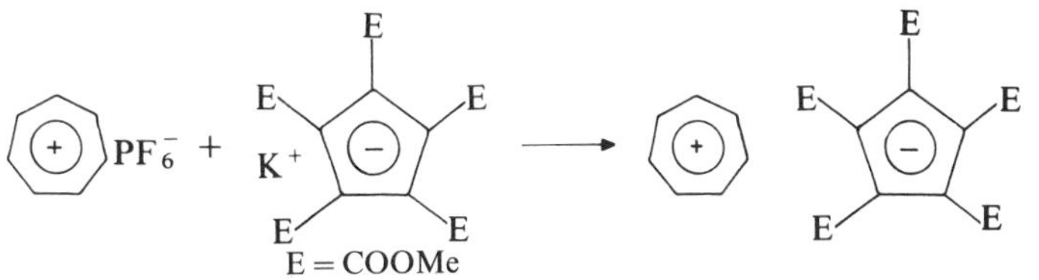

E = COOMe

This orange salt is soluble in polar solvents. An X-ray study shows it to have an infinite sandwich structure of interleaved planes of anions and cations.

Nature and Stability of Tropylium Salts

In general, tropylium salts fall into two distinct groups, depending on the nucleophilicity of their associated anions. Tropylium salts having anions of low nuceophilicity such as perchlorate or tetrafluoroborate are colourless, of high melting point and are completely dissociated. They are for the most part readily soluble in water and insoluble in organic solvents of low polarity; they can often be recrystallized from polar solvents such as acetonitrile or nitromethane. They are stable in air and sublime only with difficulty, if at all. The perchlorate is an explosive of some sensitivity which is readily detonated. It is stabilized by the presence of water.

In contrast, tropylium salts having strongly nucleophilic anions derived from weak acids, such as acetic or hydrocyanic, are only feebly dissociated into ions; C_7H_7CN exists largely as cyanocycloheptatriene. These compounds show a range of colour, sublime easily, decompose in air and are sensitive to heat and light. The halides deliquesce very readily, with decomposition. They dissolve readily in dichloromethane but decompose in solution unless rigorously protected from the atmosphere. In air they are hydrolysed, forming cycloheptatriene, tropone and benzaldehyde.

Structure

All spectroscopic data confirm the ionic structure of dissociated tropylium salts and indicate that the ring is symmetric and planar. Thus the ^{1}H-n.m.r. spectrum consists of a singlet at $\delta \sim 9.2$. This chemical shift is consistent with a delocalized cationic structure. The positive charge on the ring enhances the deshielding compared to neutral benzene, in contrast to the shielding effect of the negative charge on the cyclopentadienide anion.

Many tropylium salts having colourless anions are themselves colourless, but tropylium bromide and iodide are, respectively, yellow and red. This colour has been attributed to formation of charge-transfer complexes.

An elegant chemical demonstration of the symmetry of the tropylium ring involved the preparation of a sample of tropylium bromide having one of its seven carbon atoms labelled with ^{14}C:

· = ^{14}C-labelled carbon

This labelled salt was treated with phenyl magnesium bromide to give phenylcycloheptatriene. When this product was oxidized with permanganate, benzoic acid was formed which had a specific radioactivity just one-seventh that of the original tropylium bromide:

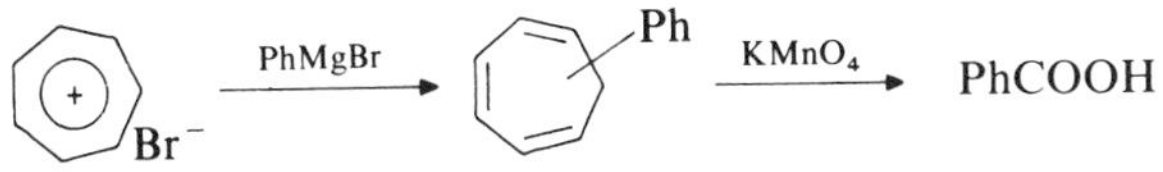

$(^{14}C^{12}C_6H_7)^+Br^-$ $^{14}C^{12}C_6H_7Ph$

Hence all the carbon atoms in the tropylium ion must be chemically equivalent.

Chemical Reactions

Not surprisingly, tropylium salts are extremely inert towards electrophiles, and usual electrophilic substitution reactions do not take place. For example, even under severe conditions in which benzene undergoes deuteriation almost instantaneously and even saturated hydrocarbons undergo rapid hydrogen/deuterium exchange, no such deuterium exchange is observed in tropylium cations.

   Despite its stability, the tropylium cation reacts very readily with nucleophiles. In many cases these reactions are governed by an equilibrium between the reactants and the resultant products. For example, in water tropylium cations are in equilibrium with cycloheptatrienol:

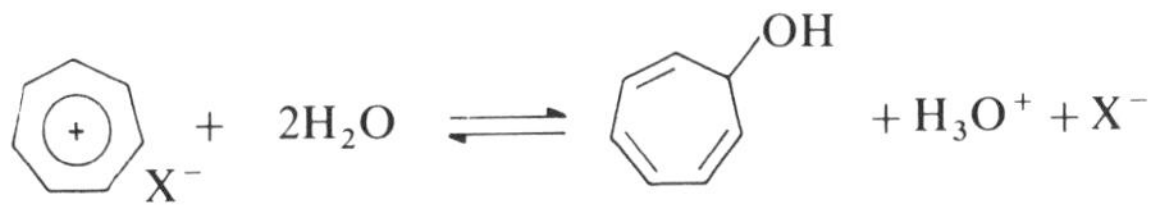

It is not possible to isolate the alcohol because it reacts with further tropylium ions to form ditropyl ether, which can be isolated:

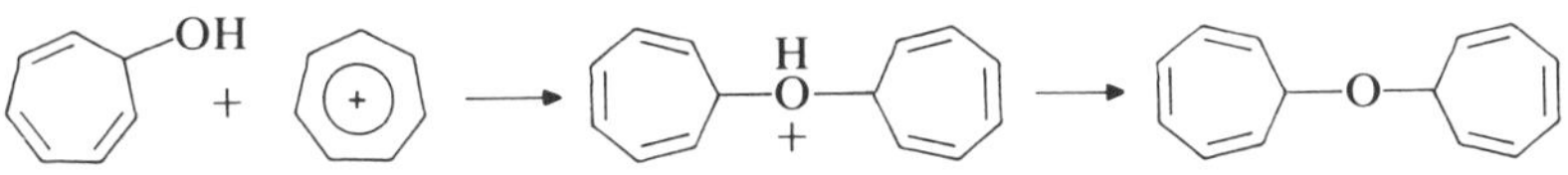

Ditropyl ether

Reaction of a tropylium salt with aqueous alkali converts it into ditropyl ether in high yield. When ditropyl ether is treated with acid the process is reversed and tropylium ions are reformed. In the presence of a very small quantity of either hydrochloric acid or tropylium ions ditropyl ether undergoes a disproportionation reaction to give cycloheptatriene and tropone, and this provides a convenient way of obtaining tropone from tropylium salts, and hence from cycloheptatriene:

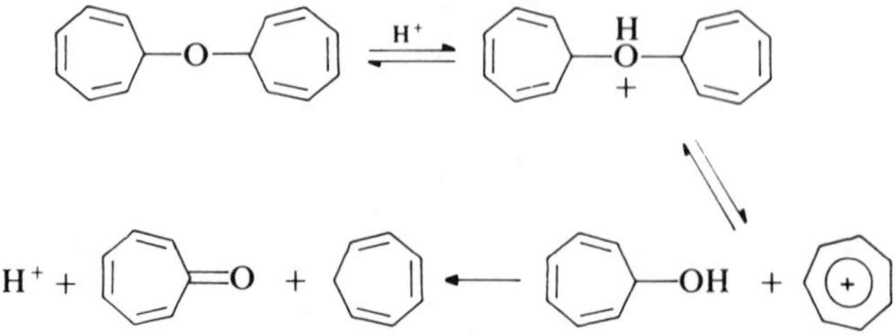

Other nucleophiles react similarly with tropylium salts. For example, sodium methoxide, potassium cyanide or arylamines give, respectively, methoxy-, cyano- and arylaminocycloheptatrienes. In some instances, as in the case of hydroxycycloheptatriene, further reaction leads to other products.

Chloro- and bromotropylium salts react with nucleophiles either by substitution of the halogen atom or by addition at a different site in the ring. Examples are as follows.

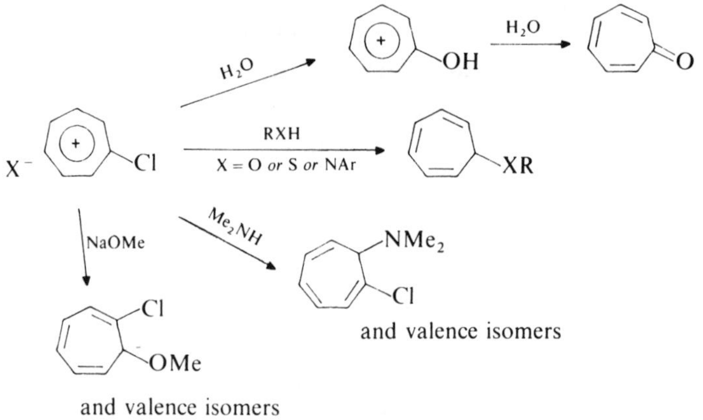

and valence isomers

Substitution of the halogen atom tends to be associated with weaker nucleophiles and is thought to be the thermodynamically controlled process.

Lithium aluminium hydride or sodium borohydride reduce tropylium salts to cycloheptatriene. Bimolecular reaction to give ditropyl can be brought about electrochemically and also by many metal powders and metal ions. Among recommended reagent are zinc powder and chromium(II), titanium(III) and vanadium(II) salts:

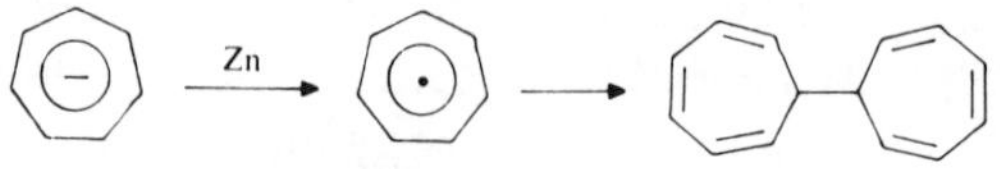

Most oxidizing agents (for example, chromic oxide or permanganate) convert tropylium salts into benzene derivatives, especially into benzaldehyde. The type of mechanism proposed is as follows:

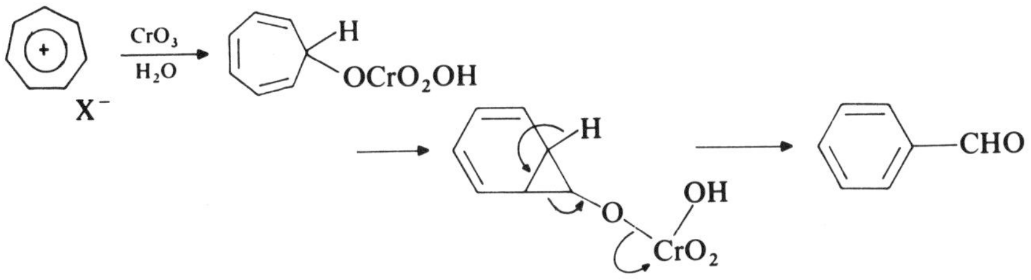

Tropone is produced when the tropylium cation is oxidized by dimethyl sulphoxide, reaction proceeding via nucleophilic attack on the cation:

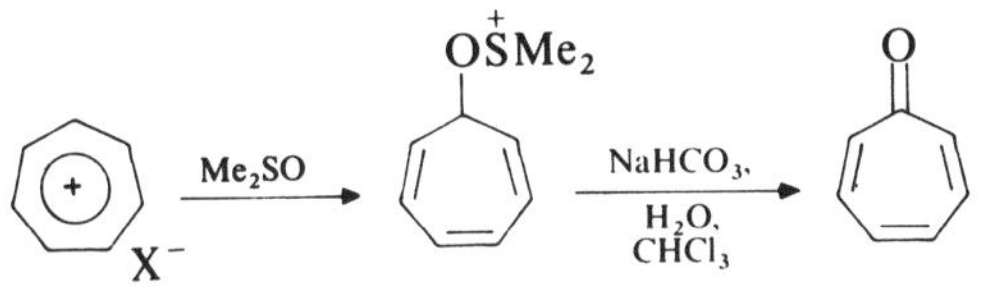

Homotropylium Salts

The concept of homoconjugation was mentioned in Chapter 5 in connection with the detailed structure of methano[10]annulene. When cyclo-octatetraene is protonated in strong acids a product is formed which spectroscopic evidence suggests to have a *homotropylium ion* structure:

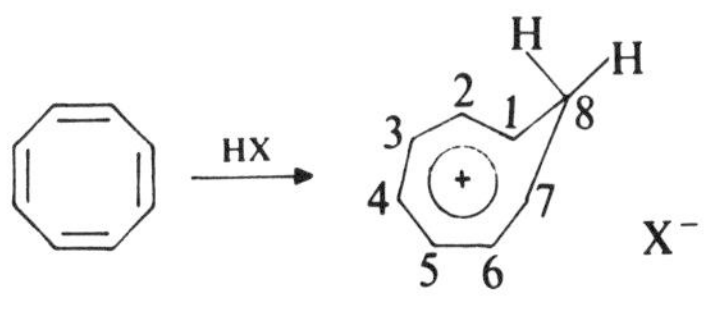

In this cation six π-electrons are delocalized over seven carbon atoms which form a coplanar ring. Six sides of this ring are made up of normal carbon–carbon bonds and the seventh takes the form of an out-of-plane methylene bridge. The assignment of this structure was based largely on the ^{1}H-n.m.r. spectrum, and in particular on the large difference in chemical shift for the two signals ($\delta 5.2, -0.6$) corresponding to the two hydrogen atoms of the bridging methylene group. This was ascribed to the fact that one of these hydrogen atoms lies over the ring, whereas the other is outside the ring, and, in consequence, the former is shielded while the latter is deshielded. Sophisticated calculations support the concept of the stable homotropylium cation and suggest that it represents an ideal example of a *homoaromatic* system. The X-ray structure of a 2-hydroxy derivative is in accord with that expected; furthermore, ^{13}C-n.m.r. spectra of this compound both in solution and in the solid state show that it has a similar structure in either state. However, the situation may be more complicated. A 1-ethoxy derivative shows a smaller difference in chemical shifts ($\delta 1.37$, 4.48) between the two methylene protons, and an X-ray structure indicates lack of conjugation across the C(1)–C(7) gap. As with bridged annulenes, the influence of substituents may be an important factor.

Boracycloheptatriene

Boracycloheptatriene is interesting in that it represents a neutral analogue of the tropylium ion, with the possibility of six π-electrons shared over seven ring atoms.

The derivative shown has been prepared. It is colourless and very sensitive to air. Its electronic spectrum shows that there is extensive conjugation of the boron with the hexatriene system and its n.m.r. spectra indicate that it is diatropic.

Aza- and Oxacycloheptatrienes

In contrast to boracycloheptatriene, the aza- and oxacycloheptatrienes (azepines and oxepins) have eight π-electrons and thus would not be expected to have stabilizing delocalized electronic systems. This is borne out by their properties. They have non-planar tub-like structures and are alkene-like. Both may undergo valence isomerization to bicyclic structures; this tendency is more marked in the case of oxepin.

Thia-azacycloheptatrienes

The trithiadiaza- and trithiatriazacycloheptatrienes represented below both have the possibility of delocalized systems of ten π-electrons:

Both have been prepared and both have considerable thermal stability, being recovered unchanged after being heated in solution in o-dichlorobenzene at 180 °C for two days.

The trithiadiaza compound is inert to protic or Lewis acids and to amines, but is rapidly destroyed by aqueous sodium hydroxide. It does not undergo cyclo-addition reactions. The trithiatriaza compound is slightly less stable kinetically and decomposes slowly in air at room temperature. Both compounds have been nitrated and brominated, at carbon, the trithiadiaza compound providing both mono- and di-substituted products. Reaction presumably proceeds via a Wheland-type intermediate, i.e.

(E^+ = electrophile)

The positive charge in the intermediate can be delocalized over all three sulphur atoms. The trithiatriaza compound is less reactive to electrophiles than is the trithiadiaza compound. Their spectra show that both compounds have delocalized electronic structures, and X-ray studies indicate that the rings are planar and symmetrical with bond lengths appropriate for delocalized systems.

It seems likely that the trithiatetra-aza compound whose formula is also given above might also be thermodynamically stable, and it would be a fully inorganic cyclic delocalized compound.

Tropones, Tropolones and Heptafulvenes

In the following sections the chemistry of compounds which, hypothetically, could be dipolar compounds with tropylium rings as the positive end of the dipoles, is considered. It is not now thought that dipolar contributions play any great role in the structure of these compounds.

Preparation of Tropones

Tropone was first prepared as long ago as 1887, by treating tropinone methiodide with alkali and the resultant product with bromine, but its structure was not recognized:

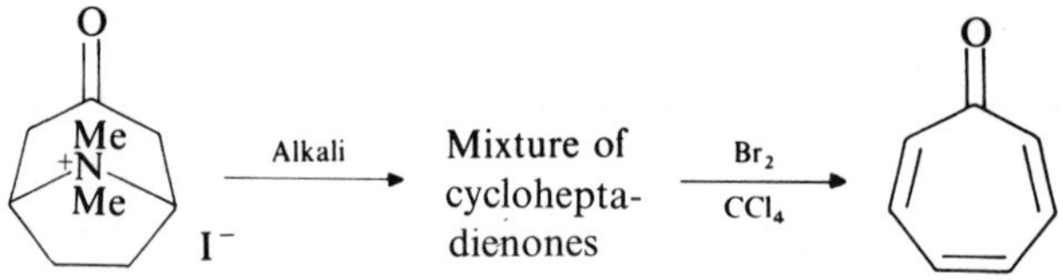

Tropinone methiodide

This work was repeated in 1953 and the product was then identified.

The most common contemporary starting materials for the preparation of tropone are cycloheptatriene or tropylium salts. One method is as follows:

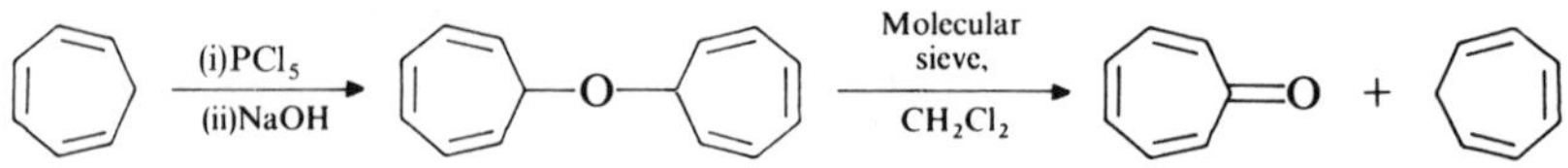

The initial step involves conversion of cycloheptatriene into a tropylium salt. Ditropyl ether may also be converted into tropone by either distilling it from acid-treated silica gel or by allowing it to stand with a trace of acid.

Other ways of obtaining tropone from tropylium salts are as follows:

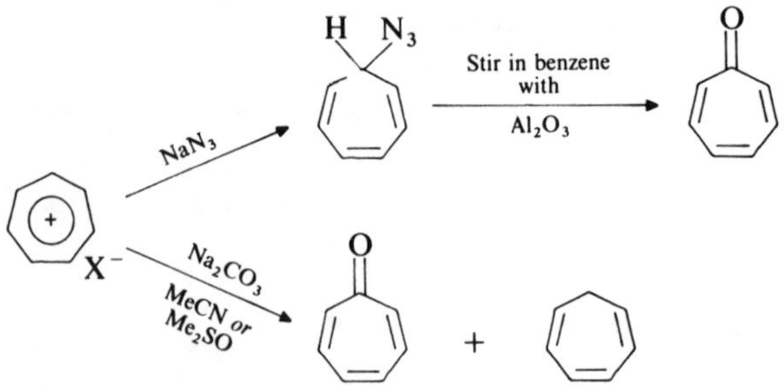

Cycloheptatriene can also be converted into tropone in good yield by anodic oxidation:

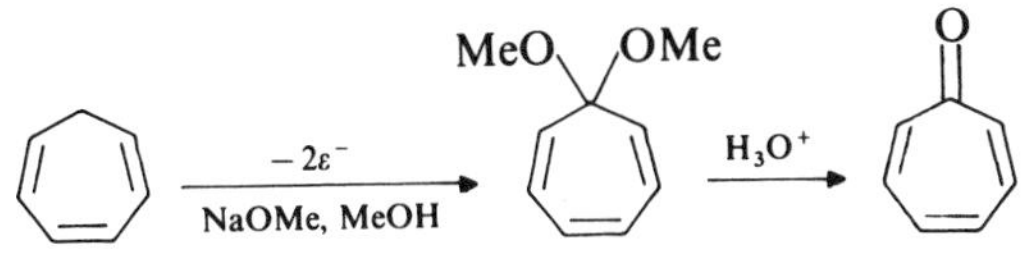

Spectra and Structure of Tropone

X-ray analysis and electron-diffraction studies show that the ring is approximately planar with bonds of markedly alternate length, about 1.36 and 1.45 Å, respectively, for the double and single bonds. There is thus little indication of cyclic delocalization of electrons in the ring. The ^{1}H-n.m.r. spectrum is in accord with this. For example, HH coupling constants across ring bonds vary from 5.2 to 13.0 Hz, and a ^{17}O-n.m.r. spectrum indicates that there is no excessive negative charge on the oxygen atom, as would be the case were a dipolar structure to make any significant contribution to the overall structure of tropone. Early work on tropone had assumed that its dipole moment (4.30D) and the carbonyl stretching frequency in its infrared spectrum (1590 cm^{-1}) indicated considerable dipolar character and delocalization of the electrons in the ring, but more complete analyses of these figures show them to be reasonable for a cyclic polyenone.

Chemistry of Tropone

Tropone is a colourless liquid which is completely miscible with water. It decolorizes permanganate and is unstable to alkali. When pyrolysed, tropone and its derivatives lose carbon monoxide and form benzene or derivatives of benzene. A norcaradiene intermediate is thought to be involved:

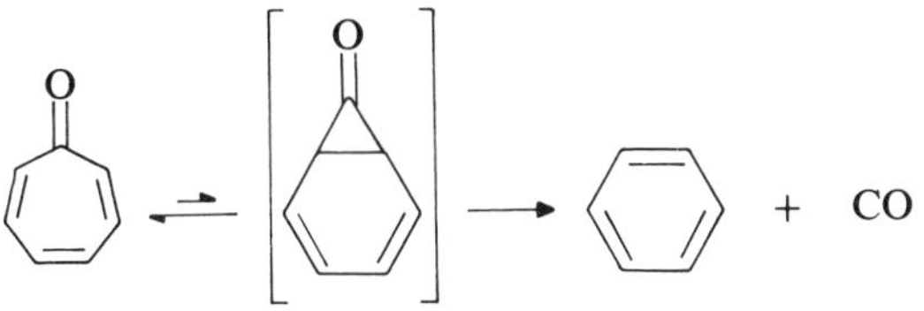

Tropone forms normal ketone derivatives, such as a semicarbazone and arylhydrazones, only under forcing conditions. It has been pointed out that there is more strain energy in a cycloheptatriene ring having one sp^3-hybridized ring atom than in a similar ring having all its atoms sp^2-hybridized, as in tropone, and this may contribute to the low carbonyl reactivity of tropone. It has a relatively high basicity for a ketone; with strong acids it gives hydroxytropylium salts:

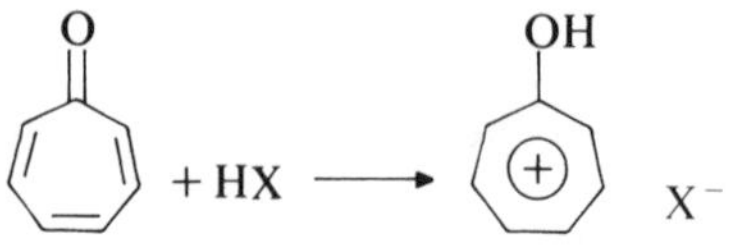

The basicity can be attributed to the unusual stability of the conjugate acid. The carbonyl group reacts similarly with acyl derivatives to give acyloxytropylium salts.

Chlorine or bromine undergo addition reactions with tropone. When they are kept, the resultant adducts tend to lose hydrogen halide, providing halogenotropones. Examples of such reactions are as follows:

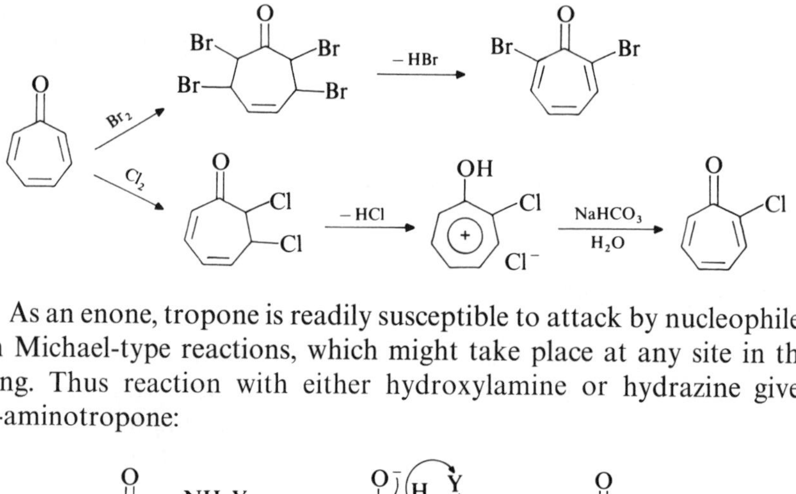

As an enone, tropone is readily susceptible to attack by nucleophiles in Michael-type reactions, which might take place at any site in the ring. Thus reaction with either hydroxylamine or hydrazine gives 2-aminotropone:

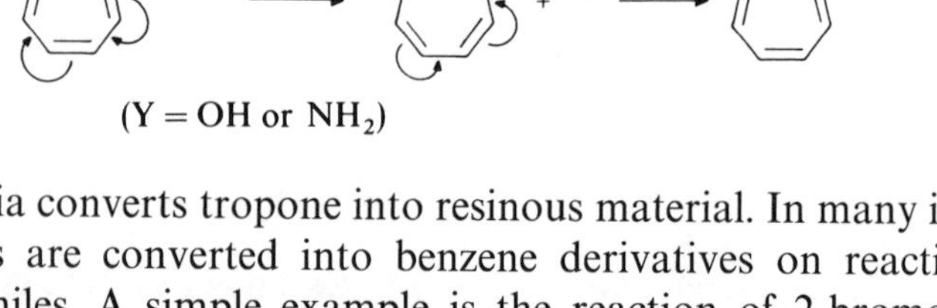

(Y = OH or NH$_2$)

Ammonia converts tropone into resinous material. In many instances tropones are converted into benzene derivatives on reaction with nucleophiles. A simple example is the reaction of 2-bromotropone with alkali, which results in the formation of a benzoate salt:

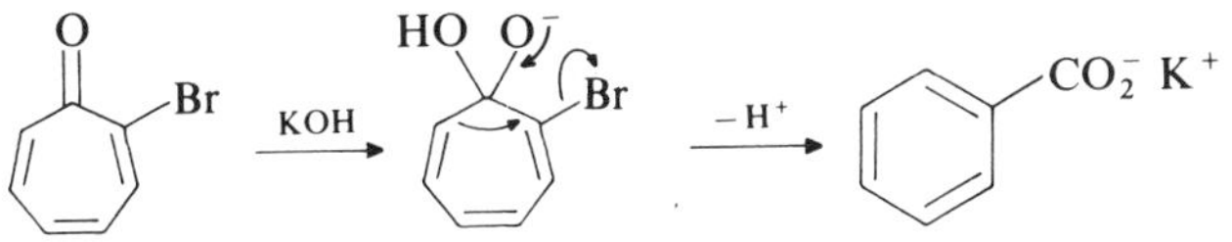

In other cases the tropone ring is retained and a substituent group is replaced by the nucleophile:

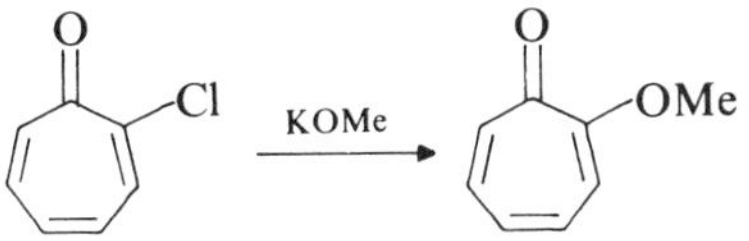

Many such reactions are not, however, straightforward nucleophilic substitution reactions but may involve nucleophilic attack at a site other than that occupied by the group which is displaced. The reactions of tropones with nucleophiles are complex, often with a number of competing pathways. A review (A. Pietra, *Acc. Chem. Research*, **12**, 132 (1979)) discusses the situation in detail.

Tropone is reactive in cyclo-addition reactions, as either a dienophile or a diene, and may react as a 2π, 4π or 6π component. With dienophiles, such as maleic anhydride, tropone reacts as a diene:

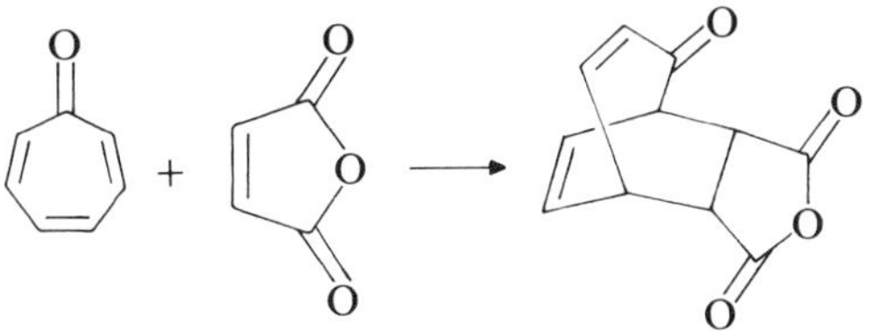

With dienes tropone commonly reacts as a 6π component: its reaction with cyclopentadiene provided the first recognized example of a thermal [6 + 4] cycloaddition reaction:

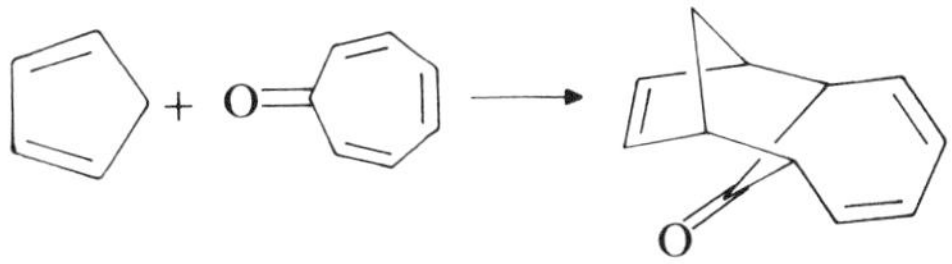

With other simple dienes tropones give a [6 + 4] adduct as major product together with a mixture of [4 + 2] adducts:

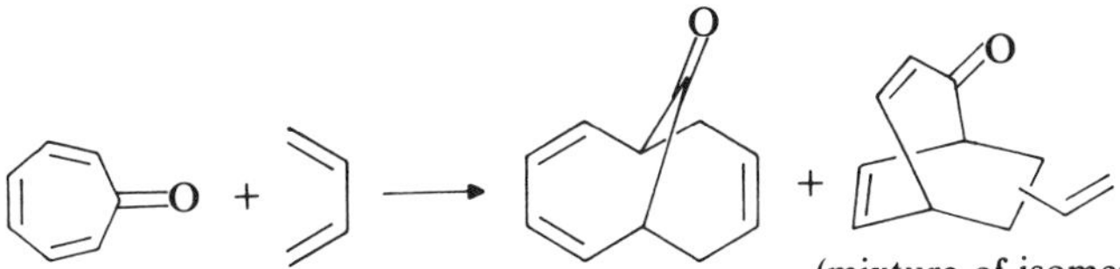

(mixture of isomers)

Tropone dimerizes readily, both thermally and photochemically.

Reaction of tropone with phosphorus pentasulphide provides thiotropone:

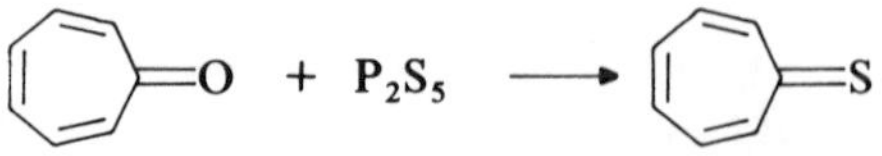

$$\text{(ring)}=O + P_2S_5 \longrightarrow \text{(ring)}=S$$

Thiotropone forms red needles. It polymerizes readily at room temperature, but can be kept at $-78\,^{\circ}C$ under nitrogen for several months. It is very labile in solution.

Preparation of Tropolones

Tropolones, or 2-hydroxytropones, excited great interest in the late 1940s as possible novel aromatic compounds. Syntheses of tropolone were first reported in 1950, starting from either benzene or cycloheptane-1,2-dione:

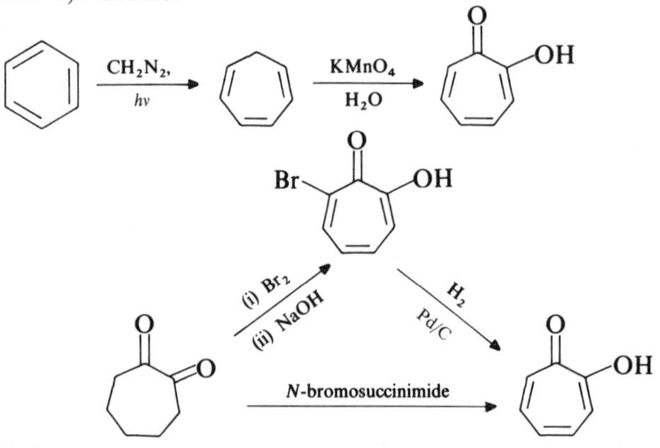

The overall yield in the route from benzene was only 1%, but the ready availability of the reactants and the elegant simplicity of the method made it none the less practical. (Cycloheptatriene is now readily available commercially.)

Tropone may be converted into tropolone via either 2-chlorotropone or 2-aminotropone:

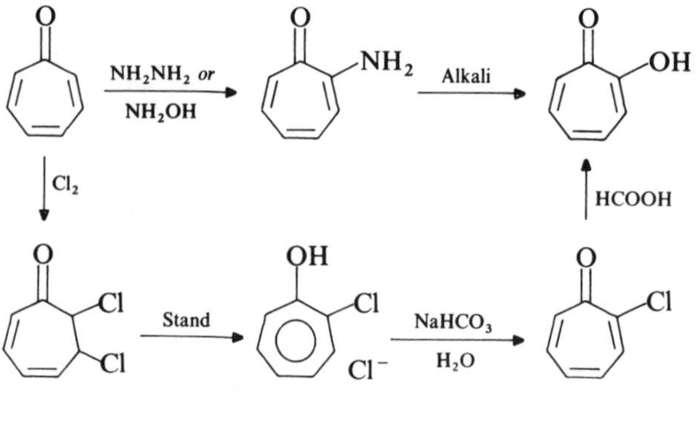

A useful method for the preparation of tropolone involves the reaction of dichloroketene, generated *in situ*, with cyclopentadiene to give a bicyclic product which can be readily transformed into tropolone:

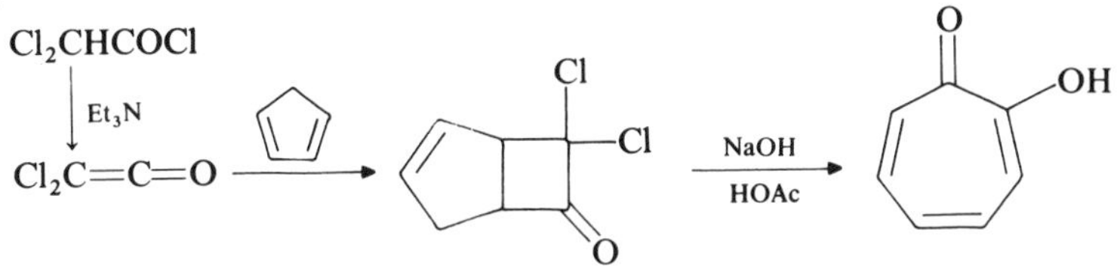

Substituted tropolones have been made similarly.

Structure of Tropolone

In the first discussions of the structure of tropolones it was pointed out that they are vinylogues of carboxylic acids and of the enol forms of 1,3-diketones:

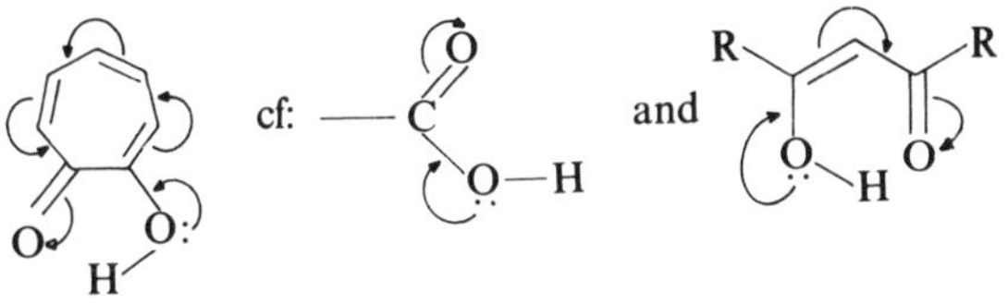

In each of these types of compounds 'push–pull' electronic interaction between an electron-donating hydroxyl group and an electron-withdrawing carbonyl one is possible. In tropolones this electronic interaction involves all the ring bonds except the C(1)—C(2) bond. Were tropolone to exist as a hydroxy derivative of a dipolar tropone structure, i.e.

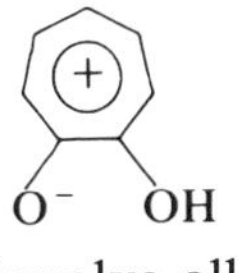

electronic interaction would involve all seven sides of the ring.

All evidence supports the first interpretation. X-ray studies show that the C(1)—C(2) bond is markedly longer than the other C—C bonds in the ring, and that there is an alternation in length in these bonds. The alternation between the bond lengths is less than in tropone because of the conjugative interaction between the hydroxy and carbonyl groups in tropolones, which involves these ring bonds.

In solution there is rapid exchange between two tautomeric forms:

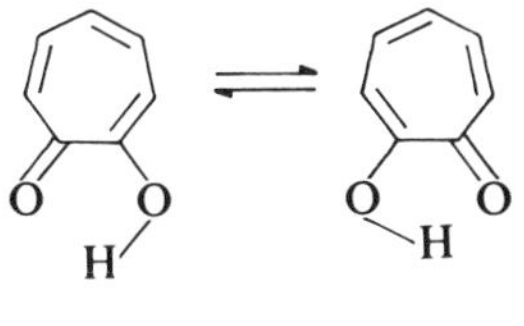

Only four signals appear in the ^{1}H-n.m.r. spectra of tropolone, even at low temperature, showing that this exchange is very rapid and that the energy barrier to interconversion of the tautomers is very low. Because of this tautomerism, separate isomers of unsymmetrically substituted tropolones (e.g. 3- and 7-methyltropolones) cannot be isolated, since they interchange too rapidly:

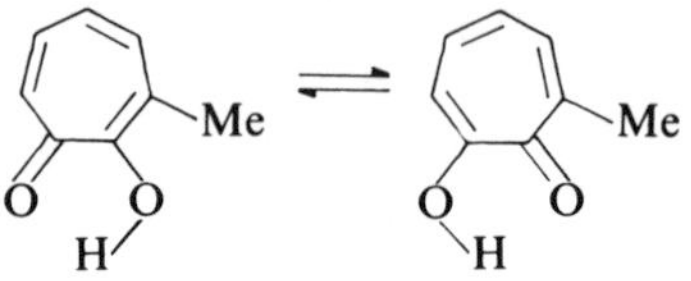

A similar situation obtains in the case of enols of β-diketones:

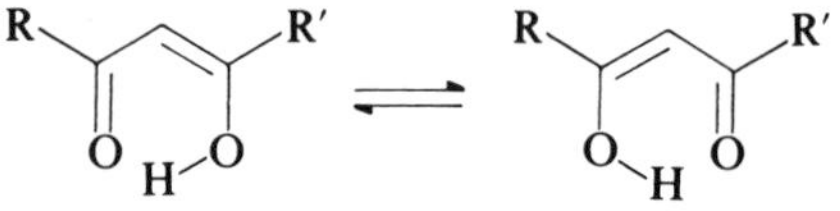

An X-ray examination of tropolone has located the hydrogen atom of the hydroxy group and showed that it was attached to one of the oxygen atoms and was not placed symmetrically between the two oxygen atoms. Furthermore, the two carbon–oxygen bonds are of different length.

Chemistry of Tropolones

Tropolones are crystalline solids which dissolve more readily in hydroxylic solvents than in ether or hydrocarbons. Many sublime readily, which has sometimes assisted in their purification.

When they were first studied, the ease with which they underwent electrophilic substitution reactions caused excitement, for this resemblance to benzenoid compounds suggested that they might be regarded as 'aromatic' compounds. However, this reactivity also shows their resemblance to β-diketones; they are vinylogues of the enol form of the latter, i.e.

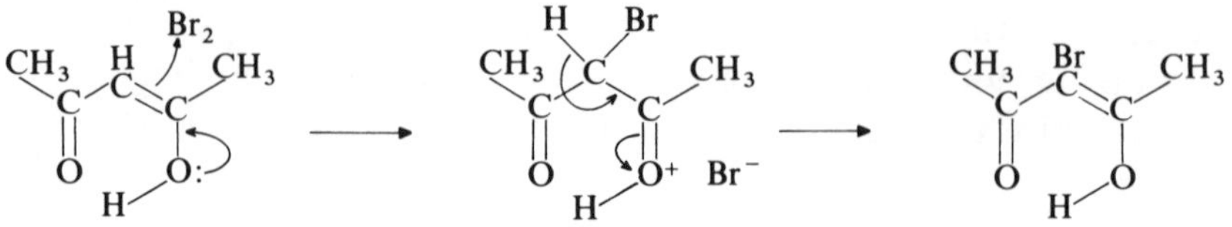

compare:

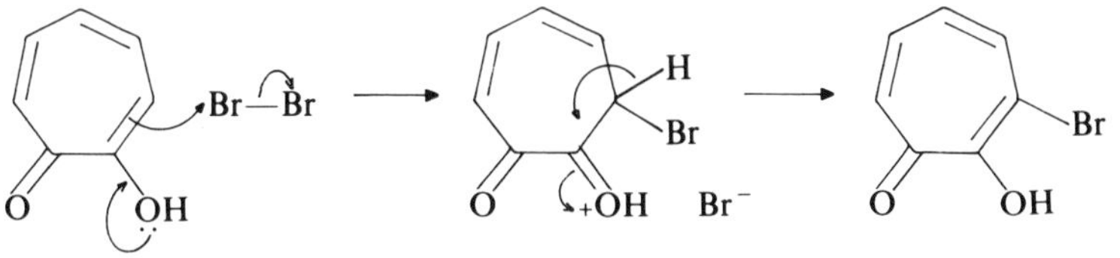

Other examples of electrophilic substitution reactions of tropolone are as follows:

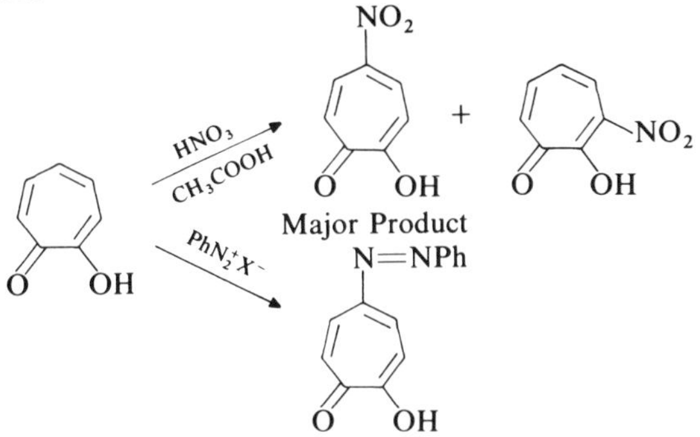

Major Product

Substitution takes place at the 3-, 5- and/or 7-positions.

Tropolone is not sulphonated by concentrated sulphuric acid; rather, protonation takes place to give a stable dihydroxytropylium ion, which is resistant to electrophilic attack:

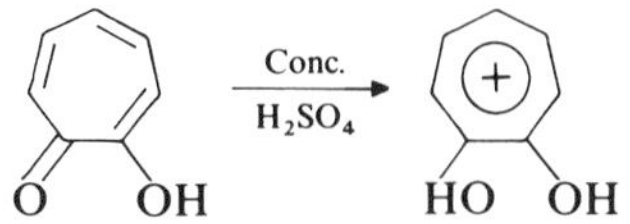

Similarly Friedel–Crafts acylation and alkylation do not take place because tropolones form complexes with the metals which are resistant to electrophilic attack.

In the reactions with electrophiles, tropolones resemble β-diketones; in other reactions they resemble carboxylic acids, of which they are also vinylogues. For example, tropolones are acidic, and there are reactions corresponding to formation of esters, acid chlorides and amides:

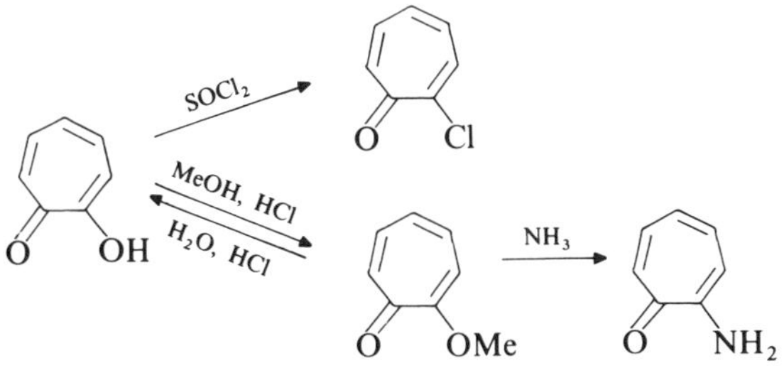

Unsymmetrically substituted tropolones give rise to two alkoxy-tropones, since tautomerism between these compounds is not possible:

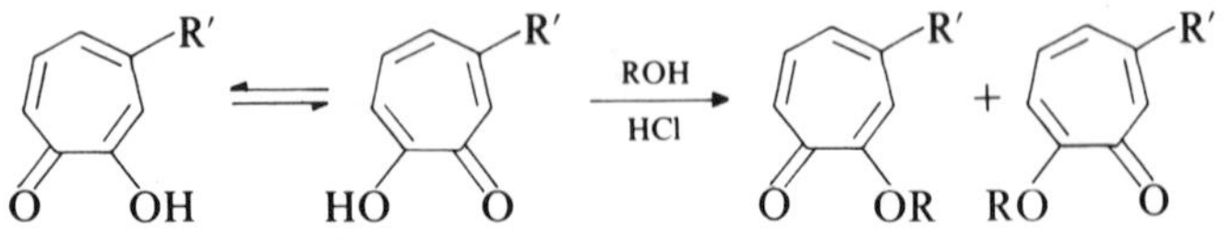

Acid chlorides or acid anhydrides react with tropolone to give O-acyltropolones, which are vinylogues of acid anhydrides, and, like acid anhydrides, they are readily hydrolysed.

As noted earlier in this chapter, 2-halogenotropones (vinylogues of acid chlorides) cannot be reconverted into tropolone by aqueous hydroxide, but instead give a salt of benzoic acid.

Heptafulvenes

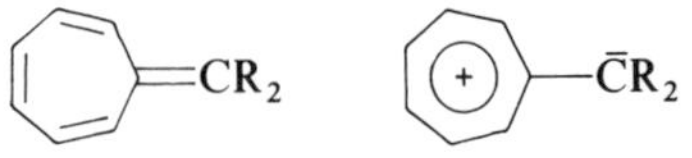

(A)   (B)

Heptafulvenes (A) are analogues of fulvenes, but in this case any contribution from a dipolar structure such as (B) would result in the exocyclic group bearing a negative charge, while a tropylium ion character developed in the ring.

As in the case of fulvenes, the parent compound (R = H) is extremely labile and polymerizes very readily, even at low temperatures. Hepta-fulvenes with electron-withdrawing groups substituted on the exocyclic carbon atom, which enable any negative charge on this atom to be delocalized, are much longer lived. Thus 8,8-dicyanoheptafulvene (R = CN) is stable up to quite high temperatures.

N.m.r. spectra indicate that heptafulvene has a largely localized double-bond structure. A dipolar structure not only increases energy by separation of charge but also increases steric strain, and it seems that the energy involved is greater than the gain in delocalization energy brought about by the presence of a tropylium ring. X-ray studies show that even in 8,8-dicyanoheptafulvene there are bonds of notably alternate length.

Summarizing Comments on the Derivatives of Cycloheptatriene

The notable point about the different derivatives of cycloheptatriene discussed in this chapter is that they represent very different types of chemical systems.

Tropylium salts are examples of stable compounds having cyclic delocalized six-π-electron systems. Tropolones are stable, but, instead of a cyclic delocalized system, the stabilization derives from a push–pull alkene system, which also makes them vinylogues of carboxylic acids and the enolic form of β-diketones. Tropones are conjugated unsaturated ketones. Heptafulvenes are polyenes. Thus the only common feature between all these sets of compounds is that they have seven-membered rings made up from sp²-hybridized carbon atoms.

Further Reading

General: F. Pietra, *Chem. Rev.*, **73**, 293 (1973).

Tropylium salts: T. Asao and M. Oda, in *Methoden der Organischen Chemie* (*Houben–Weyl*), 4th edition, Vol. 5/2c, Thieme, Stuttgart, 1985, pp. 49ff.

Homotropylium salts: R. F. Childs *et al.*, *Pure and Applied Chem.*, **58**, No. 1, 111 (1986).

Tropones and tropolones: T. Asao and M. Oda, in *Methoden der Organischen Chemie* (*Houben–Weyl*), 4th edition, Vol. 5/2c, Thieme, Stuttgart, 1985, pp. 710ff.

Reactions of substituted tropones with nucleophiles: F. Pietra, *Acc. Chem. Res.*, **12** 132 (1979).

Heptafulvenes: T. Asao and M. Oda, in *Methoden der Organischen Chemie* (*Houben–Weyl*), 4th edition, Vol. 5/2c, Thieme, Stuttgart, 1985, pp. 768ff.

See also Chapter 3 of the author's 1984 book (for details see Preface).

9

Some Other Cations and Anions Derived from Cyclic Polyenes

The ionic compounds considered in Chapters 7 and 8 have, like benzene, cyclic systems of six π-electrons. There are homologues of these compounds, with different ring sizes, having cyclic systems of $(4n + 2)$ π-electrons, with $n = 0, 2, 3$, etc.

In general, cyclic polyenes with $(4n + 1)$ ring atoms could be expected to lose a proton to provide stabilized anions with $(4n + 2)$ delocalized electrons; for example, from cyclononatetraene $(n = 2)$:

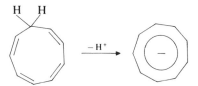

Similarly, cyclic polyenes with $(4n + 3)$ ring atoms could lose a hydride ion to provide stabilized cations. The simplest example is provided by cyclopropene $(n = 0)$:

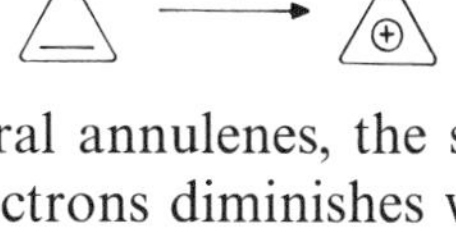

As in the case of neutral annulenes, the stabilization provided by an array of $(4n + 2)$ π-electrons diminishes with increase of ring size, but the delocalization of charge, especially of negative charge, resulting from an excess of electrons, increases the tendency for delocalization in the ring.

The most studied of these species have been those derived from cyclopropene, and they will be considered in the next section.

Following that, derivatives of cyclononatetraene will be discussed, and finally ions derived from larger-ring compounds.

Cyclopropenium Salts

It followed from Hückel's rule that the cyclopropenium cation, derived from cyclopropene by loss of a hydride ion, should be a stabilized system with two π-electrons delocalized over three ring atoms. The preparation of a cyclopropenium salt was first achieved in 1957 as follows:

The product was stable, and its n.m.r. spectra showed the equivalence of the three phenyl groups. A crystallographic examination of the corresponding perchlorate, $[C_3Ph_3]^+[ClO_4]^-$, confirmed that it was ionic, and that all the ring bonds are of equal length ($1.373 \pm 0.005 \,\text{Å}$).

The common route for the preparation of a variety of cyclopropenium salts has involved addition of a carbene to an alkyne. Unsubstituted cyclopropenium salts have been obtained by treating 3-chlorocyclopropene with a variety of Lewis acids, e.g.

Infrared and n.m.r. spectra of cyclopropenium salts are consistent with structures involving delocalization of two electrons over the ring. Thus, in the case of the unsubstituted cation, the infrared spectrum is very simple, showing only four bands, as expected for molecules of this symmetry, and the ^{1}H-n.m.r. spectrum consists of a single sharp singlet at $\delta 11.15$.

Many cyclopropenium salts are stable at room temperature, but unsubstituted salts are less easily kept. For example, the hexachloroantimonate shown above can be kept indefinitely at $-20\,^\circ$C but only for a few days at room temperature, and exposure to atmospheric moisture causes rapid blackening of the initially colourless solid. Cyclopropenium salts are insoluble in non-polar solvents but dissolve in solvents such as acetone or acetonitrile. They also dissolve in alcohols but react with them to form ethers.

Chemistry of Cyclopropenium Salts

Cyclopropenium salts are stable to acid but neutral aqueous solutions become turbid due to equilibration of the salt with a covalent alcohol:

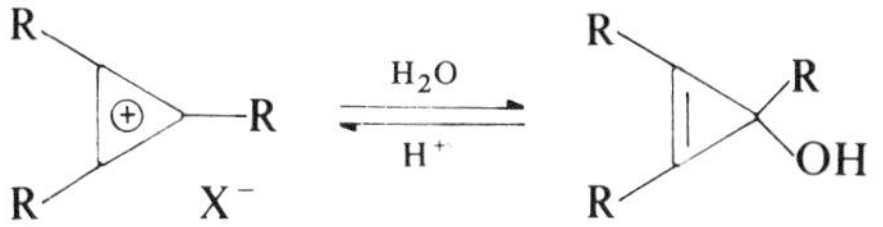

The pH value at which there is 50% ionization of the alcohol to cyclopropenium salt, designated pK_{R^+}, can be taken as a criterion of the stability of the cation. For the unsubstituted cation, $pK_{R^+} = -7.4$. Substituents which can act as electron donors provide extra stabilization; their effectiveness has been shown to be in the order $R_2N > RO > \text{n-Pr} \approx MeS > Ph > H$. Trisaminocyclopropenium salts are very stable, even to hot water.

Cyclopropenium salts are, unsurprisingly, not attacked by electrophiles. However, they are readily attacked by nucleophiles, the reaction with water mentioned above being a typical example. Other examples are:

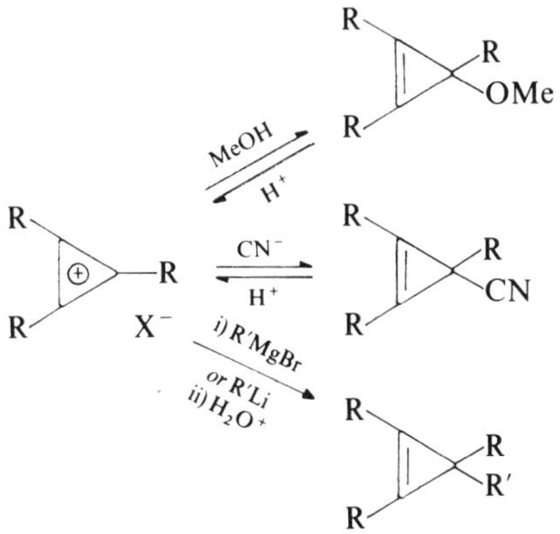

Aqueous alkali brings about ring opening, e.g.

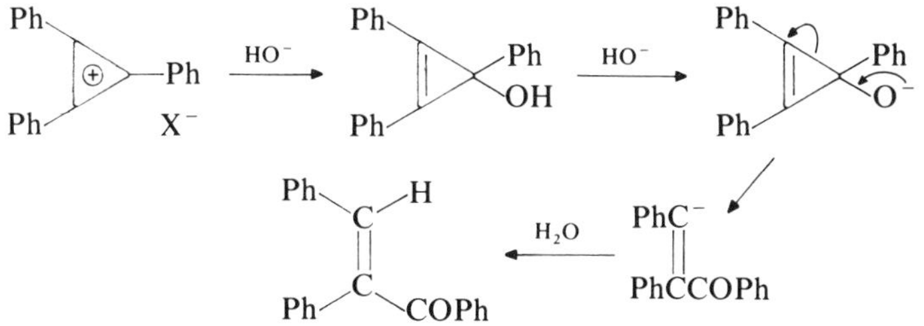

Like tropylium salts, cyclopropenium salts may undergo bimolecular reduction as well as reduction to cyclopropenes, e.g.

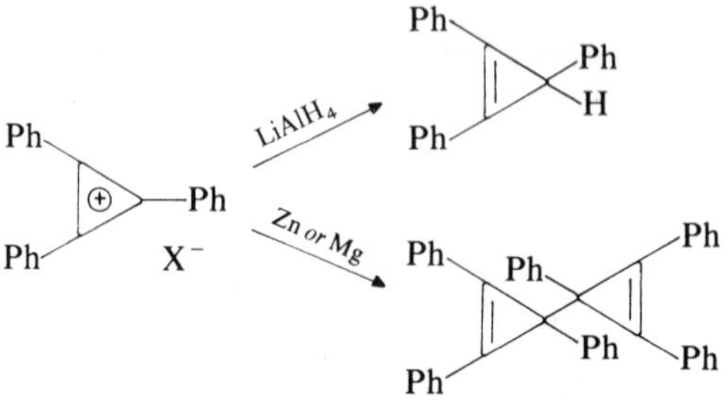

Cyclopropenones

Cyclopropenones bear the same structural relationship to cyclopropenium salts as do tropones to tropylium salts. As with tropones, it seemed possible that cyclopropenones might have a dipolar structure as an alternative to a simple ketonic one:

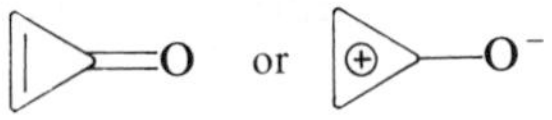

Calculations and spectroscopic data are rather inconclusive, but it appears that in the case of cyclopropenones there is a real contribution to the overall structure from a dipolar form. Thus microwave studies on cyclopropenone provide bond lengths $C=C = 1.349$ and $C-C = 1.423$, indicating some delocalization. The ^{17}O-n.m.r. spectrum shows exceptionally high shielding, also indicative of larger than usual dipolar character in the $C=O$ bond. Isotopic exchange of ^{17}O from labelled cyclopropenone by reaction with water

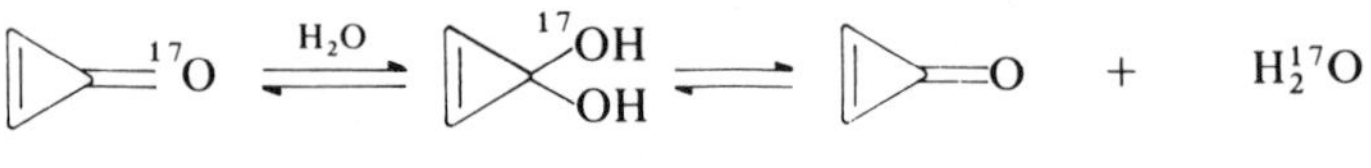

proceeds very much more slowly ($\sim \times 3000$) than for acetophenone, possibly reflecting delocalization in the three-membered ring. (Tropone reacts only $\sim \times 20$ times slower than acetophenone.) Since the stabilization associated with $(4n + 2)$-π-electron systems appears to increase with a decrease in ring size this might lead to greater delocalization in the three-membered ring and an associated greater contribution from a dipolar form. Also, in the case of cyclopropenone no extra energy is required to flatten the ring, as is required in the case of tropone.

Preparation of Cyclopropenones

One method for the preparation of cyclopropenones resembles that used for making cyclopropenium salts in utilizing the addition of a carbene to an alkyne, e.g.

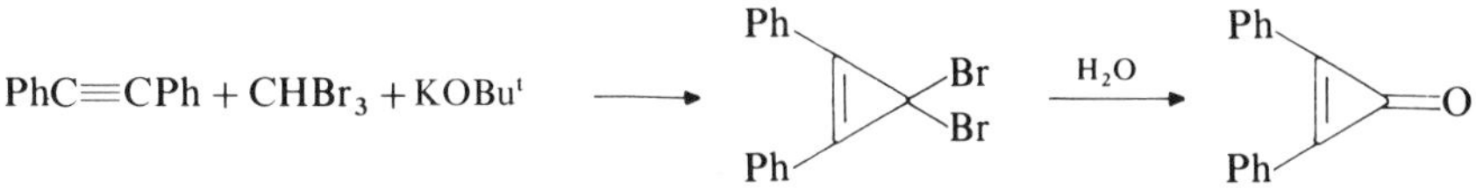

$$PhC \equiv CPh + CHBr_3 + KOBu^t \longrightarrow \text{[structure: Ph, Ph, Br, Br]} \xrightarrow{H_2O} \text{[structure: Ph, Ph, O]}$$

It is necessary to use a carbene that provides an adduct which can be converted into a carbonyl compound. An alternative method involves treatment of an α, α'-dibromoketone with a base, e.g. (*Org. Synth.*, **47**, 62 (1967)):

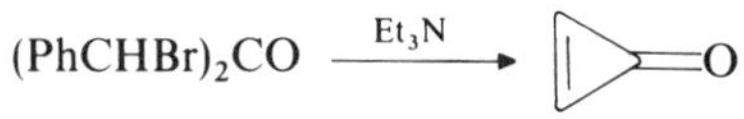

$$(PhCHBr)_2CO \xrightarrow{Et_3N} \text{[structure: O]}$$

Cyclopropenone itself has been made hydrolysis of either 3, 3-dichlorocyclopropene, 3, 3-dimethoxycyclopropene or a chlorocyclopropenium salt:

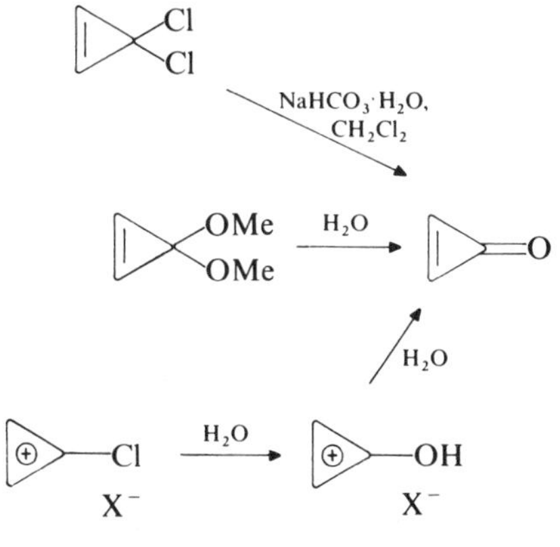

Chemistry of Cyclopropenones

In general, substituted cyclopropenones are stable compounds which melt reversibly. Cyclopropenone itself is stable only below its melting point ($\sim -28\,^\circ C$) but polymerizes at room temperature. Other cyclopropenones decompose on stronger heating, at lower temperatures giving dimers but at higher temperatures giving carbon monoxide and an alkyne:

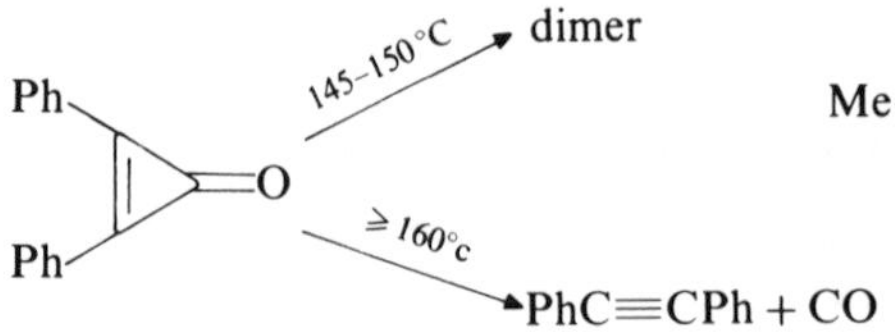

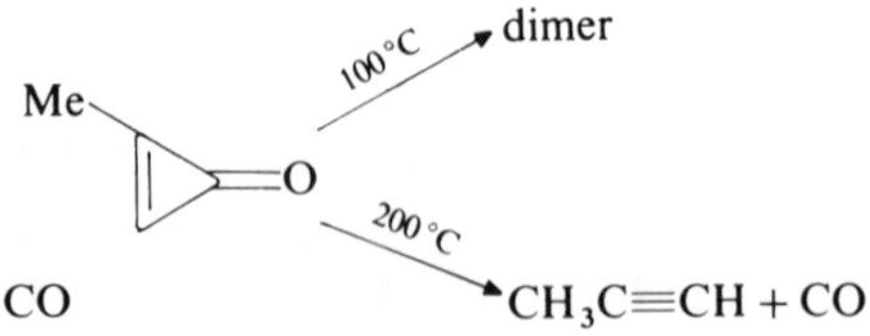

Unsubstituted cyclopropenone is stable in solution in organic solvents at room temperature. It also dissolves in water and appears not to exist in aqueous solution as cyclopropene-3,3-diol, despite the lowering of strain energy which this might provide. This is in contrast to the behaviour of cyclopropanone, which in water exists as 1,1-dihydroxycyclopropane, and may reflect stabilization in the cyclopropenone. In water cyclopropenone is slowly hydrolysed to acrylic acid. Cyclopropenone has a remarkably high boiling point (30 °C/0.45 Torr), also indicative of polar character.

Cyclopropenones are protonated in strong acids to give hydroxycyclopropenium salts:

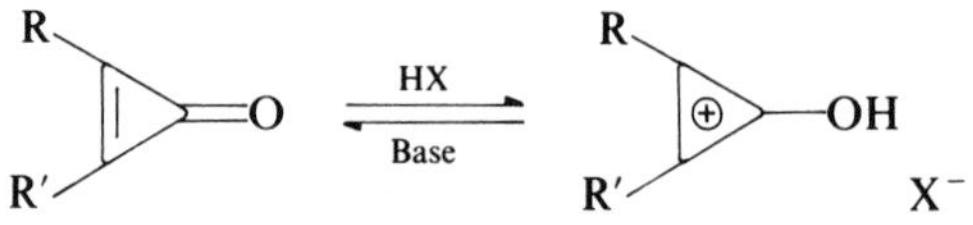

They undergo ring opening when treated with aqueous alkali:

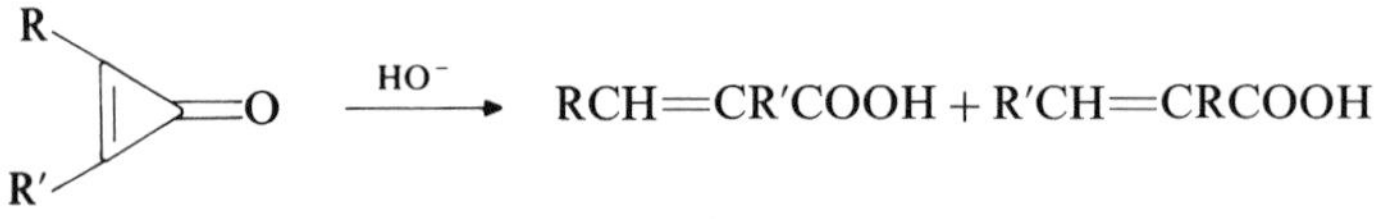

The ease with which this proceeds is very dependent on the substituents present. Thus, under conditions in which diphenylcyclopropenone is 90% destroyed in 3 min its dipropyl analogue is recovered unchanged after an hour.

Cyclopropenones react with ammonia or amines to form either alkenes or azetidines:

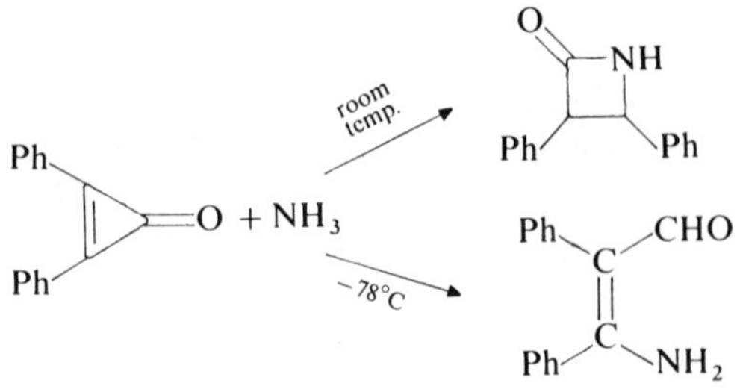

At either of these temperatures methylamine reacts to give an azetidine, while dimethylamine gives an alkene.

Cyclopropenones undergo cyclo-addition reactions across the alkene bond.

Methylenecyclopropenes or Triafulvenes

As cyclopropenones appear to have more dipolar character than tropones, so triafulvenes appear to be more dipolar and to have more delocalization in the ring than heptafulvenes. An X-ray structure determination of a dicyanodiphenyltriafulvene

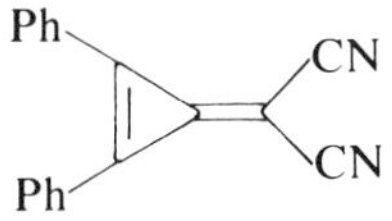

shows that the bond lengths in the three-membered ring are intermediate between those expected for either fully localized or fully delocalized structures, but suggest a significant dipolar contribution.

Triafulvene itself has been obtained in a liquid nitrogen trap from the following reaction:

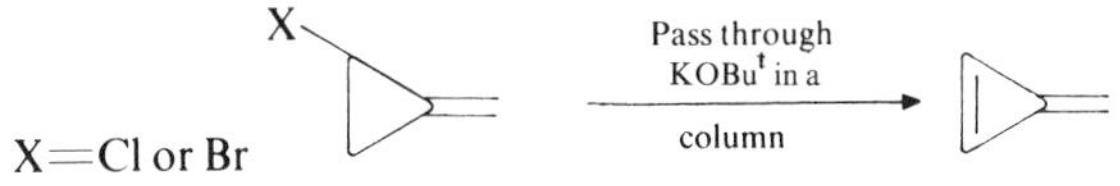

It is stable at $-100\,°C$ but decomposes at $\sim -75\,°C$. N.m.r. spectra at $-98\,°C$ suggest significant dipolar character.

Cyclopropenide Salts

The cyclopropenide anion, $[C_3H_3]^-$, has four π-electrons and should be a destabilized, or anti-aromatic, species. The difficulty in obtaining this anion testifies to this character. Reaction of cyclopropene with sodium or potassium amides in liquid ammonia results in the loss of a vinylic proton rather than of a methylene proton:

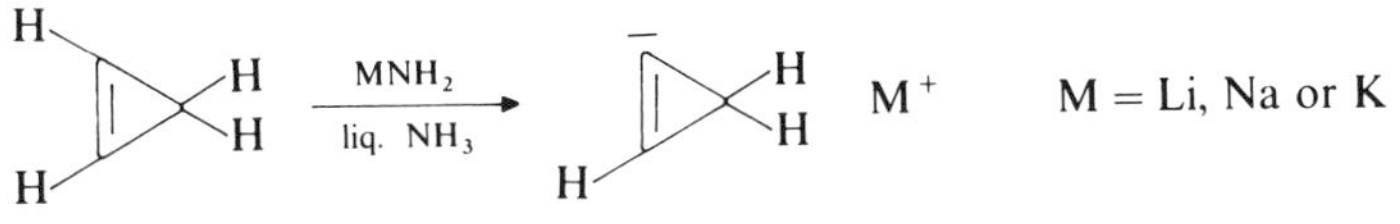

The metal ion may be partially covalently bonded to the anion. The anion which is formed is not the delocalized four-π-electron cyclopropenide anion, and its structure is such as to prevent formation of a cyclic four-π-electron system.

Cyclononatetraenide Salts

Cyclononatetraene is the next homologue of cyclopentadiene which can provide, by loss of a proton, a carbanion having a delocalized system containing $(4n + 2)$ π-electrons; in this case $n = 2$. Conversion of cyclononatetraene into a delocalized anion involves the flattening of a buckled molecule, at the same time introducing angle and conformational strain. Thus its achievement suggests that there is a compensatory factor for this introduction of strain, which must come from the stabilization introduced by setting up the delocalized ten-π-electron system.

As in the case of [10]annulene, the cyclononatetraenide, or [9]annulenide, anion might have all-*cis*, mono-*trans* or di-*trans* shapes:

The di-*trans* form is unlikely because of the severe crowding which would be imposed upon the two inward-pointing hydrogen atoms. Salts of both the all-*cis* and mono-*trans* forms have been isolated. The structures follow from their ¹H-n.m.r. spectra. The *cis* form provides a sharp singlet at $\delta\, ca\, 7$; the mono-*trans* form gives rise to two signals, at $\delta 7.27–6.4$ for the protons outside the ring and at $\delta - 3.52$ for the proton inside it. Both spectra are entirely consistent with there being diamagnetic ring currents in the rings. Since the all-*cis* form and the cyclooctatetraenide dianion have rings of similar size and ring current, the difference in chemical shifts for the protons of these two species should reflect the difference in the charge they bear; the fact that the signal in the case of the [9]annulenide is further downfield is in accord with this.

This all-*cis* form is thermodynamically more stable, and at room temperature the mono-*trans* form isomerizes to the all-*cis* form.

Variable-temperature n.m.r. studies on the mono-*trans* anion indicate that the *trans* bond migrates around the ring:

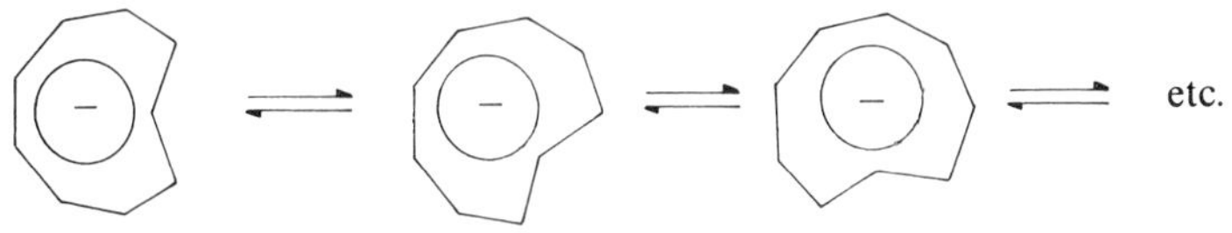

[9]Annulenide salts have usually been made from bicyclo[6.1.0]nona-triene derivatives, the latter being obtained by carbene additions to cyclooctatetraene. Thus the first preparations were as follows:

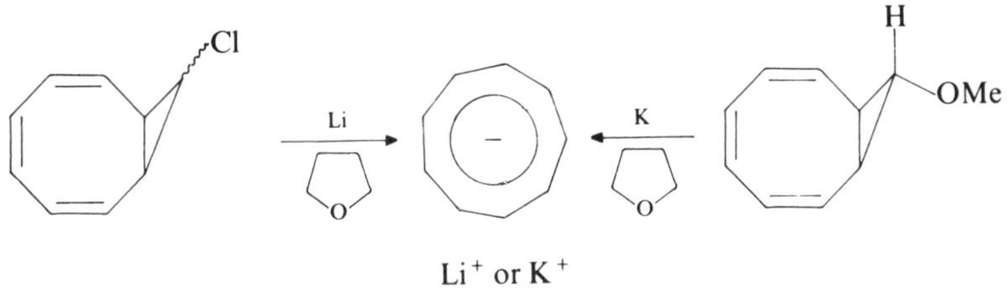

Li$^+$ or K$^+$

The ring opening of the *anti*-9-methoxy compound would be expected to proceed in a conrotatory fashion to provide the mono-*trans* ion, and when the reaction was carried out at lower temperatures this ion was indeed obtained. Both all-*cis* and mono-*trans* salts have been obtained as stable crystalline solids, but both are very sensitive to moisture and to oxygen.

A very characteristic feature of derivatives of cyclononatetraene is the ready collapse of the monocyclic system by means of an electrocyclic ring-closure reaction, providing bicyclo[4.3.0]nonatriene derivatives, i.e. dihydroindenes.

This behaviour is seen in reactions of [9]annulenide salts with electrophiles. Examples are:

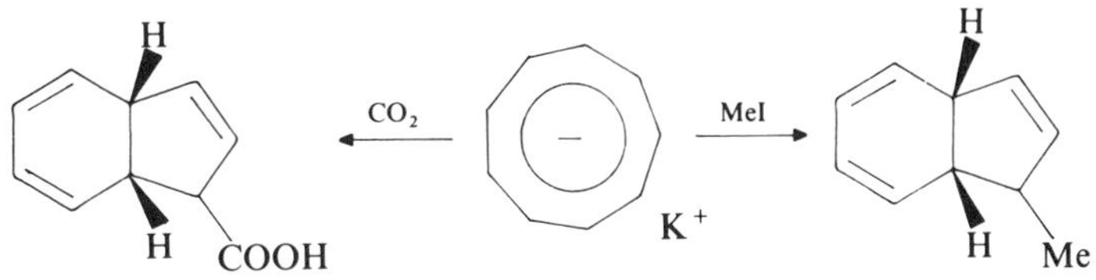

Water reacts similarly to give a mixture of dihydroindene and indene.

Bridged [9]Annulenide Ions

A methano-bridged [9]annulenide salt has been prepared. Treatment with water converts it into the tricyclic hydrocarbon from which it was initially prepared by proton abstraction:

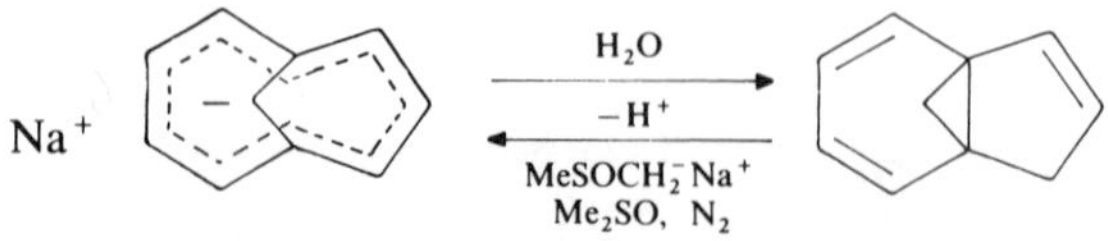

The ^{1}H-n.m.r. spectrum of this salt shows the presence of a strong diamagnetic ring current, signals for the peripheral hydrogens and the methylene group appearing at, respectively, δ 6–7 and δ − 0.45, − 0.95.

Nonafulvenes

Nonafulvene itself has only a very short lifetime, with a half-life in solution in hexane of 1 h at 10 °C. It rapidly changes into a bicyclo[4.3.0]triene:

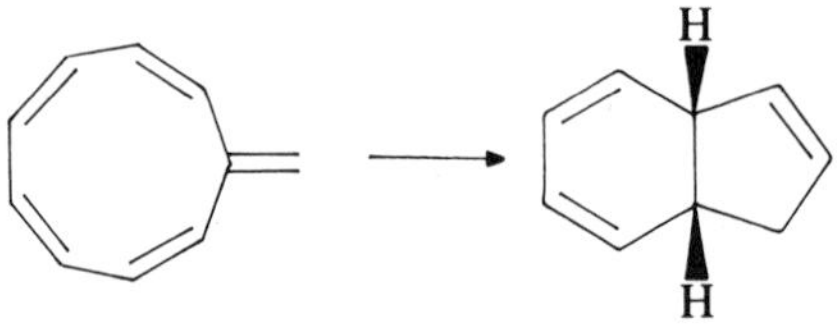

Nonafulvene itself has only a very short lifetime, with a half-life in carbon atom are, like their fulvene analogues, more stable, but are still sensitive to both air and light. Like other cyclononatetraene derivatives, they undergo ready intramolecular electrocyclic ring closure to give bicyclic isomers.

Heteronins

The heteronins are heterocyclic derivatives of cyclononatetraene, and are larger ring homologues of furan and pyrrole:

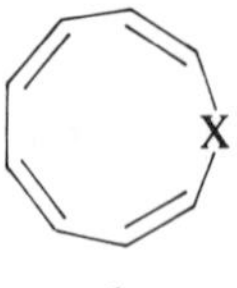

$$X = O, \ NH, \ NMe$$

They have ten π-electrons, are isoelectronic with the [9]annulenide ion, and are neutral analogues of it. Oxonin (X = O) is non-planar and atropic, with little delocalization of the π-electrons. It readily rearranges to an isomeric bicyclo[4.3.0]triene. Azonine (X = NH) and methylazonine (X = NMe) are more stable and are not isomerized up to 50 °C. They are diatropic and appear to be extensively delocalized.

Azonine provides a potassium salt which has an aza[9]annulenide anion. Like [9]annulenide salts, it is stable away from air, to which it is very sensitive, and the electrons and charge are extensively delocalized.

Other Annulenide and Annulenium Salts with Larger Rings

No simple annulenium salt with a larger ring has been described, but a bisdehydro[15]annulenium salt and a methano-bridged [11]-annulenium salt have been reported:

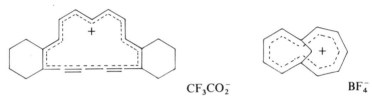

$CF_3CO_2^-$ BF_4^-

The bisdehydro[15]annulenium salt is relatively stable in solution but decomposes when solvent is removed. It is strongly diatropic (δ, outer Hs of annulenium ring 10.13, 10.31; inner Hs, − 3.13, − 3.91).

The methano[11]annulenium salt exists as stable yellow-orange needles, and is diatropic (δCH, 8.3–9.6; CH_2, − 0.3, − 1.8). It was suggested that the ^{13}C-n.m.r. spectrum could best be interpreted if the structure was regarded as a benzohomotropylium cation, but an X-ray structure determination indicated minimal orbital overlap between the 1- and 6-positions, while the peripheral bond lengths were consistent with an annulenium structure.

A 1,6-methano[11]annulenide anion has been prepared in solution and is strongly paratropic. The C(1–6) portion of the ring is non-planar; this serves both to diminish destabilizing electronic interactions of the twelve π-electrons and to relieve angle strain.

A simple [17]annulenide salt has been prepared and is diatropic (δ, outer Hs, 8.21, 9.52; inner Hs, − 7.97):

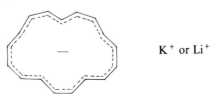

K^+ or Li^+

The acidity of cycloheptadecaoctaene appears to be similar to that of cyclopentadiene or of water. Bisdehydro[13]- and trisdehydro[17]-annulenide salts are also both diatropic.

Heterocyclic Large-ring Annulenes

Heterocyclic annulenes with odd-numbered rings are isoelectronic with annulenide anions of the same ring size. They are homologues of pyrrole, furan and thiophen. Of the latter compounds, furan least resembles benzene, and, in general, oxa-annulenes show greater polyene character than thia- and aza-annulenes.

There appears to be a general tendency for these hetera-annulenes to show less diatropic or paratropic character than all-carbon annulenes, and they have a greater tendency to be non-planar and atropic. However, great care must be taken in making any generalizations. As emphasized elsewhere in this book, many factors are at work, and there may be subtle balances between their different effects. In considering the full range of hetero-analogues of annulenes, dehydroannulenes and bridged annulenes, these factors include the rigidity or otherwise of the molecules caused by geometric factors, presence of formal triple bonds in dehydroannulenes, bridging, steric crowding, and the electron-donating ability of the hetero-atom.

[11]- and [15]hetera-annulenes might be paratropic. A thia[11]-annulene and the bridged thia- and oxa[11]annulenes shown below are all atropic and have buckled rings:

[11]   [15]   [15]

A variety of bridged and dehydro[15]annulenes have been reported, including the examples shown above, and are generally paratropic, as is an azabisdehydro[19]annulene.

A number of aza-, oxa- and thia[13]- and -[17]annulenes, dehydro-annulenes and bridged annulenes have been prepared. In the majority of cases the aza and thia compounds are diatropic, while the oxa compounds are atropic. In the case of the aza-annulenes, the presence of an electron-withdrawing group, such as an acyl or ester group, on the nitrogen atom makes the compound atropic; presumably the substituent group competes successfully with the ring for the interaction of the lone pair of electrons associated with the nitrogen atom:

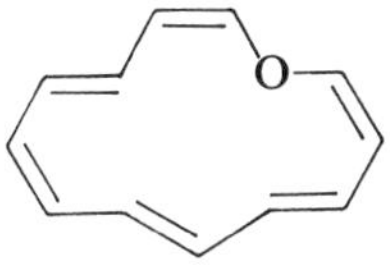

[13] atropic

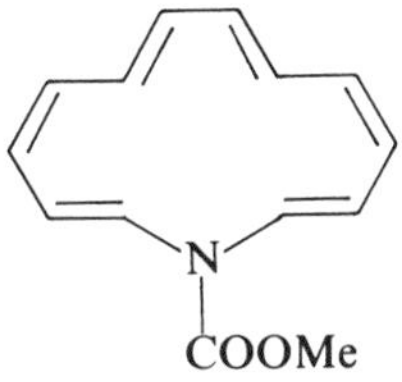

[13] atropic

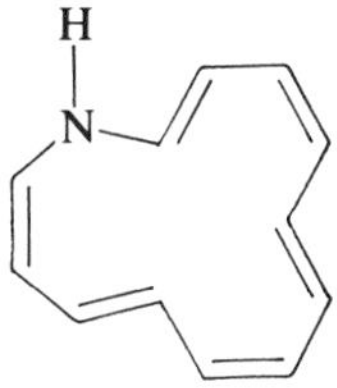

[13] diatropic

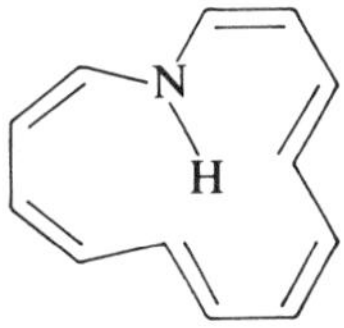

[13] diatropic

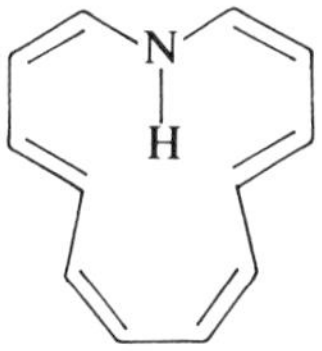

[13] diatropic

These hetera-annulenes may take up different geometries, as illustrated by the examples above. The three aza[13]annulenes are separate entities. It is interesting to note the difference in chemical shift of the n.m.r. signals for the N—H proton; depending upon whether the hydrogen is sited outside or within the ring, the signals are at $\delta 7.20$ or 0.6.

The [13]- and [17]aza-annulenes can form salts by removal of the hydrogen attached to nitrogen. The anions so formed are strongly diatropic, delocalization of charge fortifying the delocalization of the 14 or 18 π-electrons. This may be seen from the chemical shifts of the inner C—H atoms:

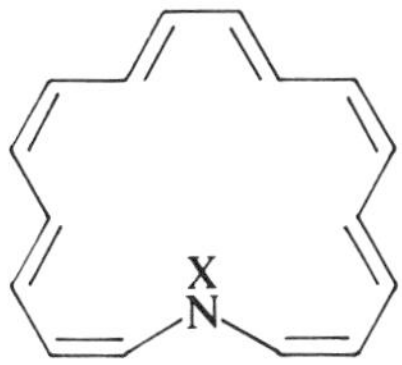

δ, inner C—H, X = H, 2.5–0.5

X = negative charge, −4.9

[17]

A bridged aza[21]annulene is known and is also diatropic.

Further Reading

Cyclopropenium salts and cyclopropenones: A. W. Krebs, *Angew. Chem.*, 77, 10 (1965); *Angew. Chem. Int. Edn Engl.*, 4, 10 (1965); I. A. D'yakonov and R. R. Kostikov, *Russ. Chem. Rev.*, 557 (1967); R. Breslow, *Angew. Chem.*, 80, 573 (1968); *Angew. Chem. Int. Edn Engl.*, 7, 565 (1968).

Cyclopropenium salts: S. V. Krivun, O. F. Alferova and S. V. Sayapina, *Russ. Chem. Rev. (Engl.)*, 43, 835 (1974).

Cyclopropenones: K. T. Potts and J. S. Baum, *Chem. Rev.*, 74, 189 (1974).

Heteronins and other larger-ring hetera-annulenes: A. G. Anastassiou and H. S. Kasmai, *Acc. Chem. Res.*, 5, 281 (1972); A. G. Anastassiou, *Acc. Chem. Res.*, 9, 453 (1976); *Pure Appl. Chem.*, 44, 691 (1975); *Adv. Het. Chem.*, 23, 55 (1978).

Articles in *Methoden der Organischen Chemie (Houben–Weyl)*, 4th edition, Vol. 5/2c, Thieme, Stuttgart, 1985, on cyclopropenium salts: G. Becker and H. Blaschke, pp. 1ff.; on cyclopropenones: G. Becker, pp. 467ff.; on cyclopropenide salts and cyclopropenyl radicals: G. Becker, pp. 38ff.; on cyclononatetraenide salts: H. J. Lindner, pp. 94ff.; on nonafulvenes: G. Becker and H. J. Lindner, pp. 787 ff.

10

Fulvalenes: Compounds with Two Separate Conjugated Rings Linked Through a Double Bond

The name fulvalene was first applied to the compound made up from two five-membered rings:

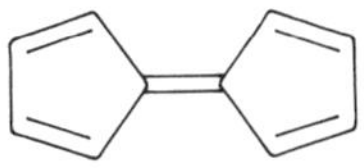

More recently *fulvalene* has been adapted to describe the whole family of this type of compound, irrespective of the size of the rings. If both rings are of the same size a single prefix is added denoting the size of the rings, i.e. fulvalene is now called pentafulvalene. If the two rings have different sizes two prefixes must be used. Thus the fulvalene in which a five-membered ring is linked to a seven-membered ring is called pentaheptafulvalene, although in this particular case the older, trivial, name sesquifulvalene is frequently employed. Symmetric fulvalenes will be considered first, and then fulvalenes with rings of different sizes.

Pentafulvalene (Fulvalene)

Pentafulvalene can be made from sodium cyclopentadienide by the following steps:

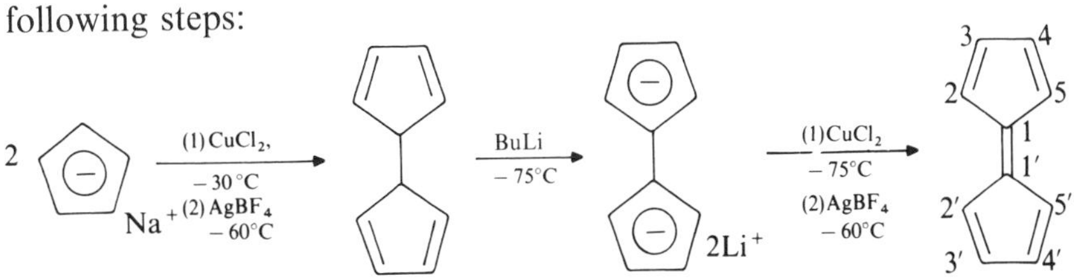

Pentafulvalene itself polymerizes very readily. Its π-electron system is more localized than in fulvenes and there are markedly alternating bond lengths. Its ^{1}H- and ^{13}C-n.m.r. spectra closely resemble those of fulvene. Introduction of bulky substituents increases the longevity of pentafulvalenes; thus the 2,3,4,5-tetraphenyl derivative is stable thermally in air. Pentafulvalenes react both as dienes and as dienophiles.

Tetrathiafulvalene derivatives have in recent years excited much interest because of the important role that they can play as electron donors in charge-transfer complex salts, which show high electrical conductivity:

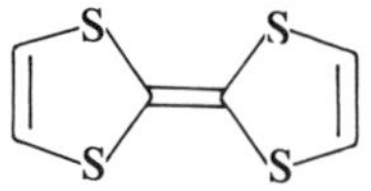

It is thought that the high polarizability and D_{2h} symmetry of this system is of fundamental importance in achieving this conductivity.

Heptafulvalene and Nonafulvalene

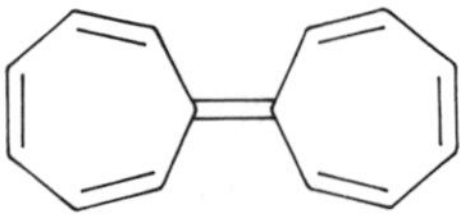

Heptafulvalene forms red-purple crystals which are unstable to light. The rings are buckled and the molecule, viewed edge-on, is S-shaped. It has localized single and double bonds.

It undergoes interesting [14 + 2] cyclo-addition reactions, which, as predicted by the Woodward–Hoffmann rules, are antarafacial, e.g.

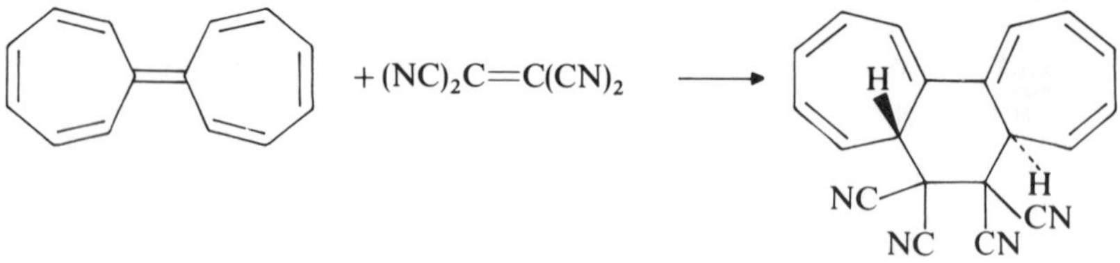

Nonafulvalene is extremely reactive and, like other cyclononatetraene derivatives, easily undergoes valence isomerism:

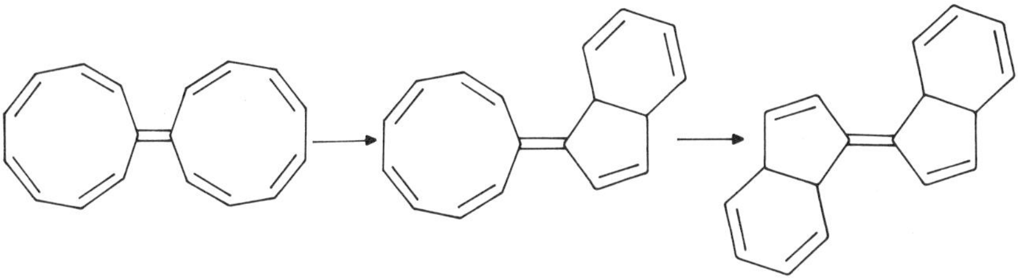

Fulvalenes with Rings of Different Sizes

If the rings of a fulvalene are of different sizes, then there may be some tendency for the electrons to be associated preferentially with one of them. This is especially the case if one ring has $(4n + 1)$ atoms and the other has $(4n + 3)$ atoms. For example, *sesquifulvalene* could have a dipolar contribution to its overall structure, which might be a hybrid of the following polyene and dipolar extremes:

In fact, fulvalenes of this sort are in general satisfactorily represented by non-polar localized structures. Only triapentafulvalene, or *calicene*, derivatives give some indication of dipolar character. In most cases the energy required to separate the charges is greater than the stabilization energy attained by setting up $(4n + 2)$ systems of π-electrons in the separate rings. The greater polarity of calicene compared with sesquifulvalene is reminiscent of the greater polarity of cyclopropenone compared to tropone.

Sesquifulvalenes or Pentaheptafulvalenes

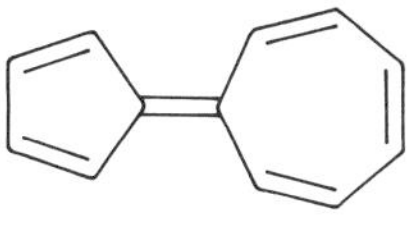

Sesquifulvalene is best represented as a polyene. Formation of a dipolar structure requires not only separation of charge but also the overcoming of strain involved in making the seven-membered ring planar. The energy required is not sufficiently compensated for by the gain of delocalization energy which is achieved thereby. All the spectroscopic evidence favours this structure for the ground state, although it seems that there is a low-lying excited state which is dipolar.

Sesquifulvalene forms red-brown leaflets and is moderately stable in solution, but it polymerizes very readily when heated or in the presence of acid. Some sesquifulvalenes with electron-withdrawing substituents on the five-membered ring appear to have more dipolar character.

Calicenes or Triapentafulvalenes

The trivial name is derived from the pictorial resemblance of the formula to a cup or chalice (Latin, *calix*, a cup):

Calicene itself has not been described but a number of substituted derivatives are known which are thermally stable. Spectroscopic evidence suggests some dipolar character, but X-ray crystallographic studies on some of them clearly show the alternation of single and double bonds, as implied by the above formula, so that delocalization of electrons and dipolar character are not a major factor.

Introduction of electron-withdrawing substituents in the five-membered ring might be expected to increase the dipolarity:

The dicyanocalicene shown has a dipole moment of 14.3D, and its electronic and n.m.r. spectra also indicate appreciable separation of charge. It and its 3,4-dicyano isomer are both thermally stable and are insensitive to light or air. The diester shown undergoes a range of electrophilic substitution reactions at the 2-position. Variable-temperature n.m.r. studies imply that there is reduced double-bond character in the bond linking the rings, which is again consistent with a measure of dipolar character.

An interesting bis-calicene has been prepared:

This compound, which is orange, is stable. N.m.r. spectra show that it is completely planar, and suggest that there is an unexpectedly large tetrapolar contribution to the ground state. There is also the possibility of complete peripheral conjugation, but 16 π-electrons would be involved, and this type of interaction appears to make no contribution.

Fulvalenes with Larger Rings

A number of fulvalenes have been prepared which have five- or seven-membered rings joined to larger rings, which are in most cases either dehydro or methylene-bridged rings. All are effectively non-polar, although there appears to be a slight tendency for electrons to move into a five-membered ring or out of a seven-membered ring, irrespective of the size of the rings to which they are attached.

This is the case, for example, in pentanonafulvalene, despite the fact that either a five- or nine-membered ring might be expected to have some tendency to attain anionoid character. However, this tendency should also be greater for the smaller ring. The nine-membered ring in pentanonafulvalene is non-planar. The compound is thermally unstable and highly reactive, and has a half-life at $-15\,°C$ of only $\sim 35\,min$.

Further Reading

M. Neuenschwander, *Pure Appl. Chem.*, **58**, No. 1, 55 (1986).
Various articles in *Methoden der Organischen Chemie* (*Houben–Weyl*), 4th edition, Vol. 5/2c, Thieme, Stuttgart, 1985, on pentafulvalenes: G. Becker, pp. 685ff.: heptafulvalenes: T. Asao and M. Oda, pp. 781ff.; sesquifulvalenes: G. Becker, pp. 697ff.; calicenes: G. Becker, pp. 479ff.; triaheptafulvalenes: G. Becker, pp. 485ff.
See also Chapter 8 of the author's 1984 book (for details see Preface).

11
Fused Ring Compounds: Benzo Derivatives

The best-known example of a benzo-annulene is naphthalene, which is, structurally, benzobenzene.

Naphthalene

Naphthalene closely resembles benzene. It is stable, undergoes electrophilic substitution reactions, has an electronic spectrum similar to that of benzene and in its ^{1}H- and ^{13}C-n.m.r. spectra the chemical shifts resemble those associated with benzene.

As a substituted benzene it will, however, as discussed in Chapter 1, show some differences in properties from benzene. At a very simple level it provides two signals in its ^{1}H-n.m.r. spectrum [$\delta 7.73(4H)$, $7.38(4H)$], since the hydrogen atoms are in two different environments, four of them (α, or 1, 4, 5, 8) being adjacent to the second ring and four (β, or 2, 3, 6, 7) being remote from the other ring:

The n.m.r. spectra and other properties indicate that there is delocalization of the π-electrons, but what form does this delocalization take? The Kekulé-type formula shows that there are ten π-electrons. We have seen in Chapter 1 that it is therefore undersirable to represent naphthalene by a formula with two inscribed circles, i.e.

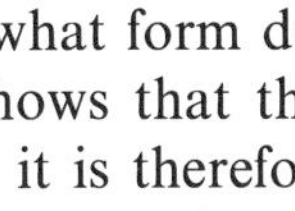

This immediately implies the presence of twelve π-electrons. Since only ten π-electrons are available there cannot be simultaneously two sextets.

Since there is energetic gain in the setting up of delocalized groups of $(4n + 2)$ π-electrons, possible ways of achieving this must be considered. There are three possibilities, two, (A, B), with a sextet in one of the rings and two alkene bonds in the other ring, and one, (C), involving delocalization of all ten π-electrons around the periphery:

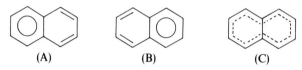

(A)   (B)   (C)

Since a sextet of π-electrons is more stabilizing than a decet, it is likely that (A) and (B) may be favoured over (C). But how can any decision be made between (A) and (B)?

At this naive level of argument the best conclusion is that the overall structure is best represented as a mean of (A) and (B), which may be regarded rather as two Kekulé forms, with perhaps a small contribution also of (C). Although (A) and (B) are more favourable than (C) the difference may not be so large as to allow (C) to be completely ignored.

If this interpretation is accepted, the central bond will have a length and bond order resembling that of benzene; if (C) were a major contributant this bond would be longer than a bond in benzene and more resemble a single bond. Of the other bonds, those linking α to β carbon atoms should have more double-bond character and be shorter than the ring bonds in benzene. This is in accord with the experimental evidence:

Experimental bond lengths (Å) 1.404 1.393
in naphthalene:
(cf. benzene, 1.395 Å) 1.365 1.424

In this text a Kekulé-type formula will be used throughout to represent naphthalene. As has been emphasized earlier, all formulae are fundamentally symbols; no single simple symbol can satisfactorily represent the structure of naphthalene, but a Kekulé-type formula at least signifies the presence of ten π-electrons and the possible benzenoid and alkenoid structures of the two six-membered rings.

Naphthalene undergoes electrophilic substitution more readily than benzene does. This can be correlated with the fact that, in the case of benzene, formation of the intermediate involves loss of the delocalized system of six π-electrons, whereas in the case of naphthalene a sextet is retained in the one ring in this step:

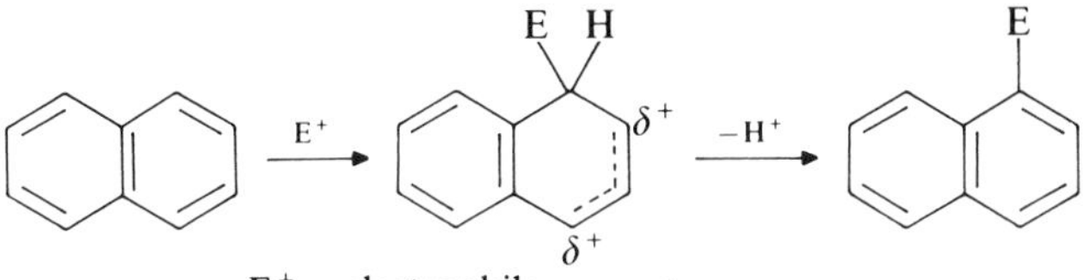

E^+ = electrophile

The preferred site for electrophilic substitution is the α-position. In this case the intermediate contains a benzene-like ring and an allyl cation. In the case of β-substitution the allyl cation is replaced by a localized cationic centre:

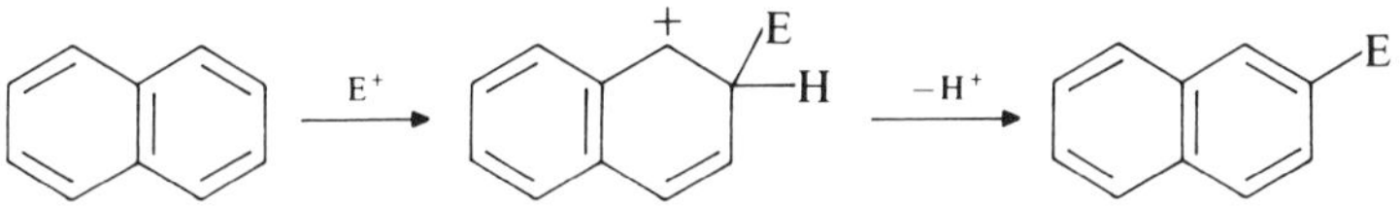

(In either α- or β-substitution there may also be some interaction between the sextet of electrons in the one ring and the adjacent positive charge in the other.)

In the case of sulphonation the α-sulphonic acid is formed preferentially at lower temperatures and the β-sulphonic acid at higher temperatures. This is a consequence of kinetic control at lower temperatures but thermodynamic control at higher temperatures. The latter favours the β-product because of steric hindrance between the 1-sulphonic acid group and the 8-position.

One ring of naphthalene is also more easily oxidized or reduced than is benzene; the resultant benzene ring only reacts under much more severe conditions:

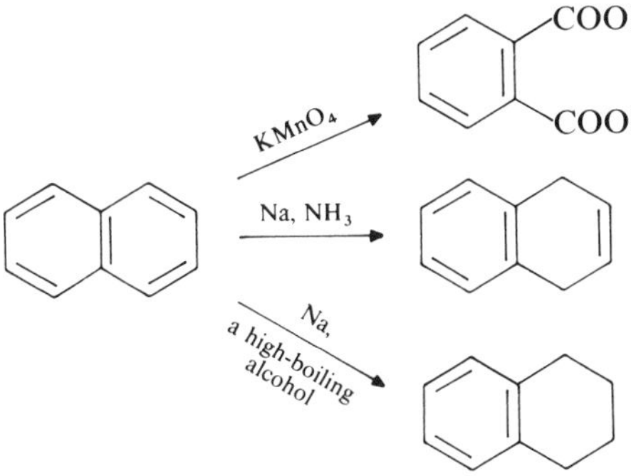

Phenanthrene

When a third six-membered ring is fused onto naphthalene, two different isomers are possible, the linear isomer anthracene and the angular isomer phenanthrene:

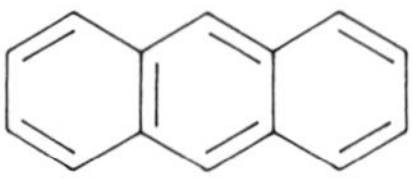

Anthracene 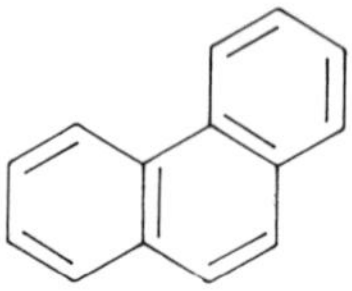   Phenanthrene

The Kekulé-type formula which best represents phenanthrene is that shown. It can be most economically considered as two rings each having a sextet of π-electrons, linked together both by a single bond and an ethylene bridge (the latter acting in effect as a largely isolated double bond).

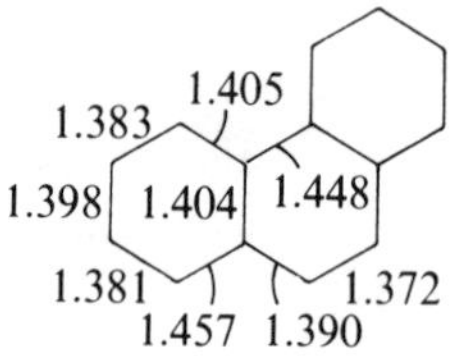

Bond lengths (Å) in phenanthrene

Although this formula provides a reasonable guide, it is, of course, oversimplified, and there must also be lesser contributions from other arrangements of electrons. Using inscribed-circle formulae to specify the location of sextets of electrons, contributing forms are as follows:

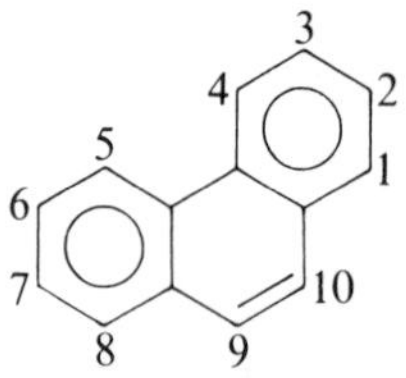

(A) (major) 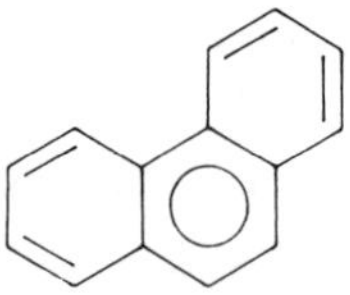   (B) (lesser)

(A) will be favoured over (B) because it provides two delocalized sextets of electrons. A peripheral circuit of 14 π-electrons is also a possible contributing structure, but probably plays only a small part since less stabilization is provided by a 14-electron circuit.

There are five chemically different sites in the molecule, represented by sites 1 and 8, 2 and 7, 3 and 6, 4 and 5, and 9 and 10. Not surprisingly, most reactions take place at positions 9 and 10, because two sextets of electrons are thereby retained:

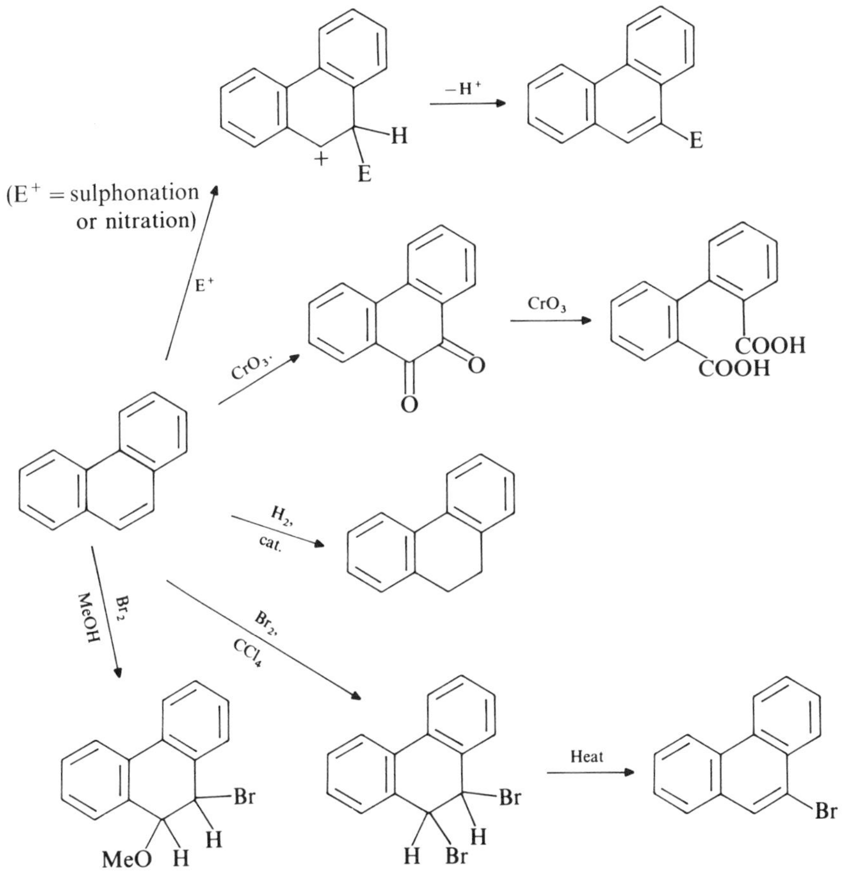

This applies both to electrophilic substitution reactions and to addition reactions. At higher temperatures, sulphonation gives 2- and 3-sulphonic acids, formed under thermodynamic control. Phenanthrene does not react with maleic anhydride in a cyclo-addition reaction.

Anthracene

Anthracene, like phenanthrene, has three fused benzene rings and 14 π-electrons, but, however these π-electrons are arranged, it is only possible to have one circuit of six π-electrons. Again using inscribed circles to represent such sextets, the possibilities are:

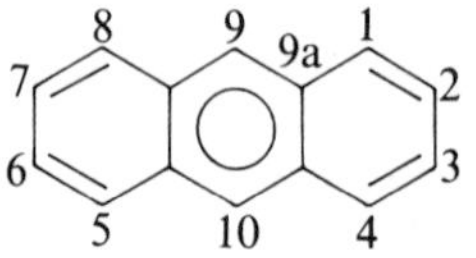

(A) 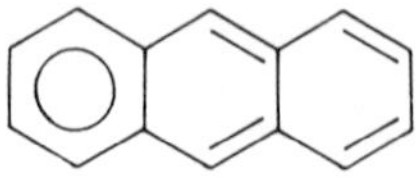   (B)

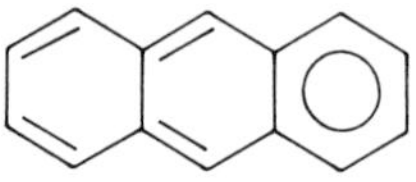

(C) 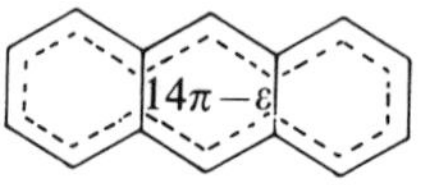   (D)

A peripheral circuit of 14 π-electrons, as represented in formula (D), is also possible, but, because of the lesser stabilization provided by these larger circuits of electrons, it is probably not a major contributor to the overall structure. Again it may be noted that formulae such as

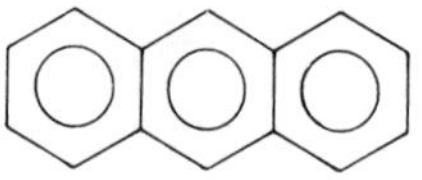

(X)

(X) are not only inadequate but incorrect.

Structure (A) probably represents the major contributing form. This is in agreement with the measured bond lengths in anthracene (Å):

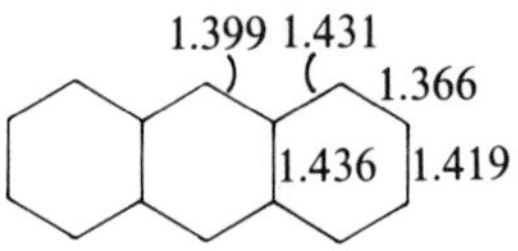

As expected, the 1,2-bond is shorter and the 1,9a- and 2,3-bonds are longer than the 9,9a-bond. The length of the bridging bonds reflects contributions from structures (B) and (C) and also possibly some contribution from (D). The inability of anthracene to sustain more than one sextet of π-electrons (whereas isomeric phenanthrene can sustain two sextets simultaneously) is reflected in the greater thermodynamic stability of the latter compound.

In the present discussion a Kekulé-type formula is used as the symbol for anthracene:

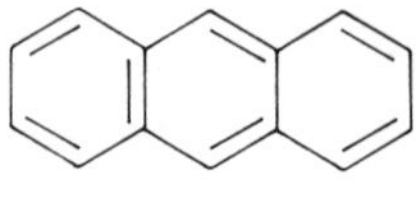

Perhaps surprisingly, since delocalization is most marked in the central ring, most reactions of anthracene involve the 9- and 10-positions. For example, halogenation, acetylation and nitration take place at these sites. The key to these reactions is that in the intermediate step a species having two delocalized sextets of π-electrons is generated; if the transition state leading to the intermediate resembles the product, the activation energy barrier will be low:

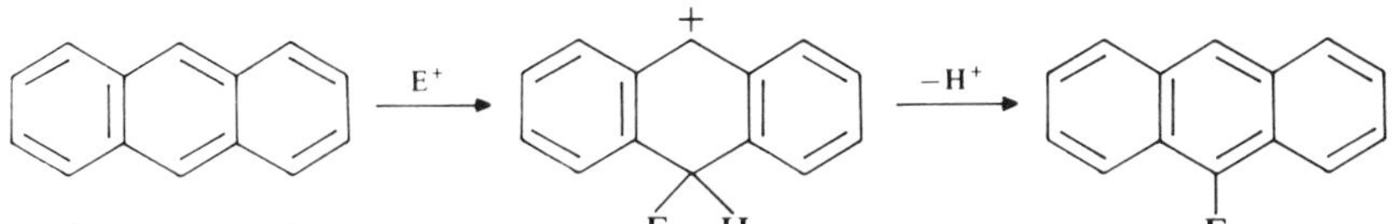

E^+ = electrophile

A further contributing factor is the extra stabilization provided to the intermediate cation by interaction with the adjacent benzene rings.

As in the case of naphthalene, sulphonation proceeds differently, to give 1- and 2-sulphonic acids. It seems probable that steric hindrance is again involved, directing reaction away from the 9,10-positions.

The reactivity of the 9,10-positions, leading to products with two benzene rings, is also shown by the ease of reduction of anthracene, and its very ready participation in cyclo-addition reactions. The latter reactivity is in contrast to that of phenanthrene:

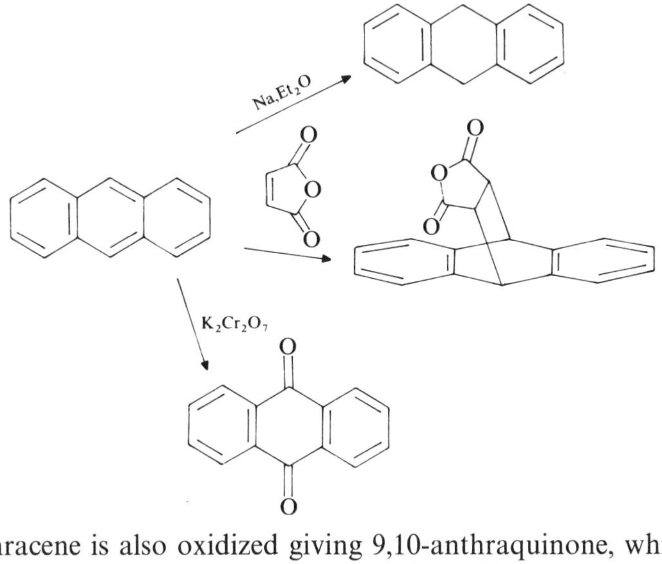

Anthracene is also oxidized giving 9,10-anthraquinone, which is a very stable quinone, showing no true quinonoid character. Rather it resembles a not very reactive ketone, such as benzophenone. This is also borne out by its molecular structure, which shows that in effect it consists of two benzene rings linked by two carbonyl groups:

Bond lengths in anthraquinone (Å)

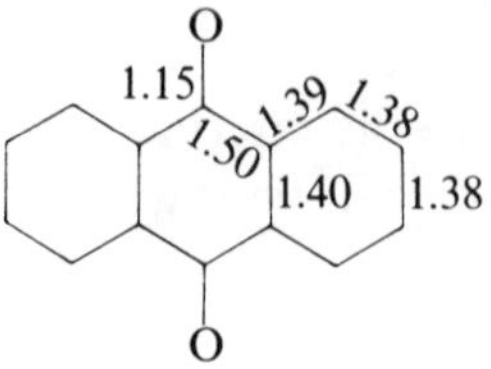

Cf. standard
C=O 1.20
sp²-C—CO   1.48
C—C(benzene) 1.395

Linear Acenes

Acenes are molecules with a series of benzene rings fused together. They are named according to the number of rings present; thus for compounds with four, five, six or seven rings fused together the names are, respectively, tetracene, pentacene, hexacene, heptacene. In the linear acenes, the colour increases and the stability decreases, the greater the number of rings. Anthracene is colourless, tetracene orange-yellow, pentacene blue-violet and hexacene green. Pentacene and hexacene are sensitive to air and light, and heptacene is sufficiently reactive never to have been obtained as a pure sample.

In all cases, as in anthracene, only one ring of a linear acene can attain a sextet of π-electrons. Hence tetracene is probably a hybrid of structures (A–D):

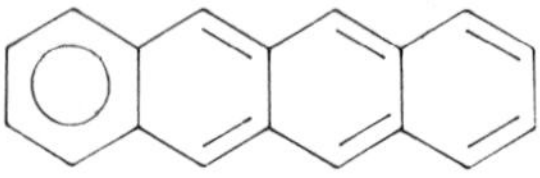

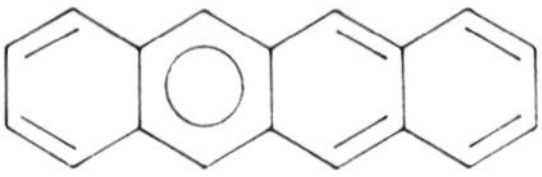

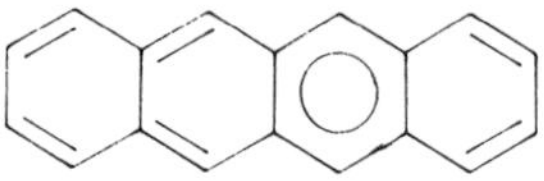

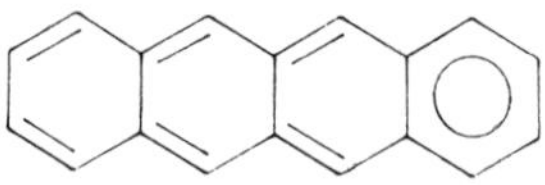

(B) and (C) probably make the largest contributions to the overall structure. It is not likely that the possible peripheral conjugation of the 18 π-electrons makes any appreciable contribution, but there may be a contribution from structure (E). Bond fixation is particularly marked in the outer rings.

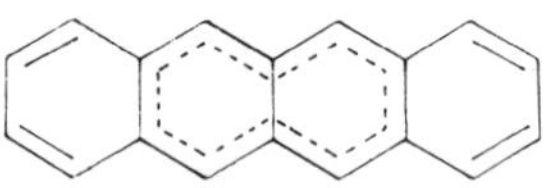

Similar considerations apply to pentacene, wherein a delocalized sextet may occupy any of the five rings, but the largest contribution is likely to come from the structure with the sextet in the innermost ring:

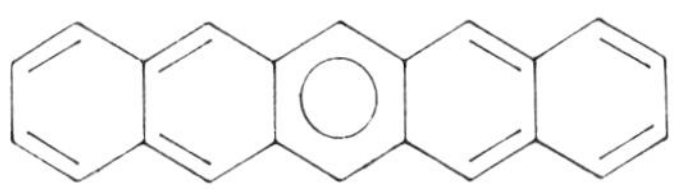

The observed bond lengths in these acenes conform with these structures.

In consequence, these acenes have largely polyalkene structures. If reactions take place across *meso* positions, i.e. *p*-positions in the same ring, more stable systems may be generated, since the products contain two smaller but better stabilized acene units. Thus linear acenes readily undergo cyclo-addition reactions. For example, tetracene reacts with dienophiles to give products which have benzene and naphthalene ring systems in the molecules, e.g.

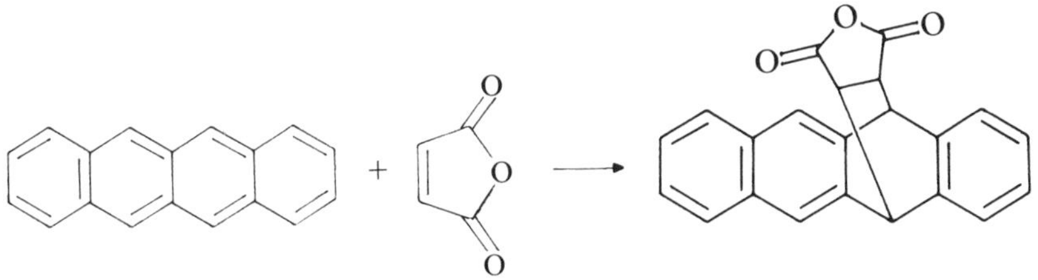

Similarly, in the presence of air and light, they readily add oxygen across *meso* positions, and this provides a danger in their preparation, for these oxides may explode if heated.

For similar reasons, phenols derived from linear acenes very readily tautomerize to give ketones, e.g.

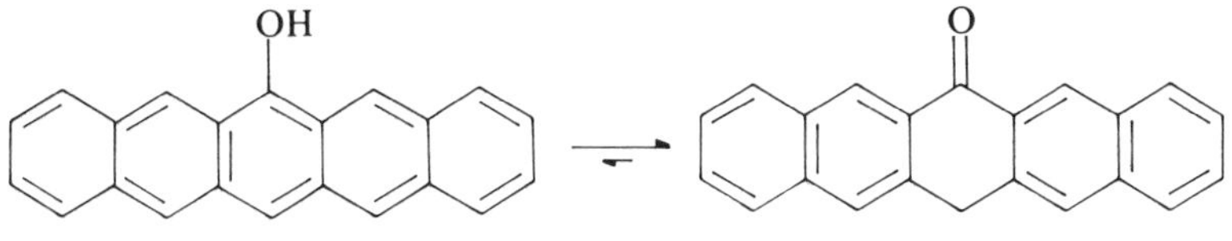

Perhaps more strikingly, methylacenes are readily converted into methylene isomers:

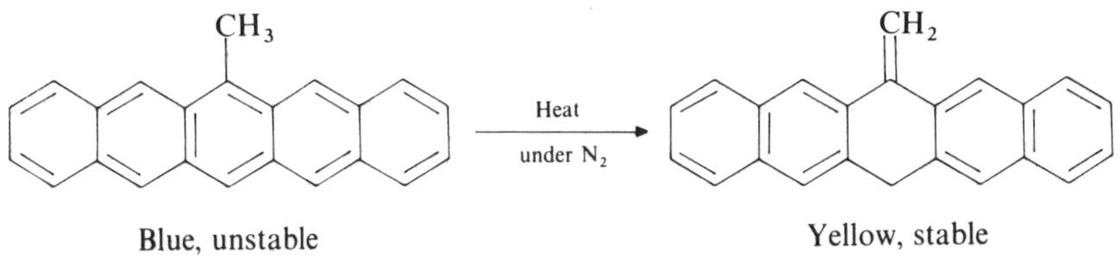

Blue, unstable   Yellow, stable

Angular Acenes

The most studied angular acenes are the *helicenes*:

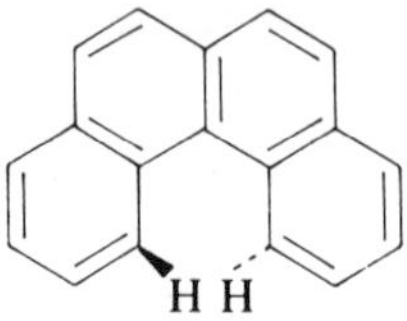

Benzo[*c*]phenanthrene
or [4]helicene

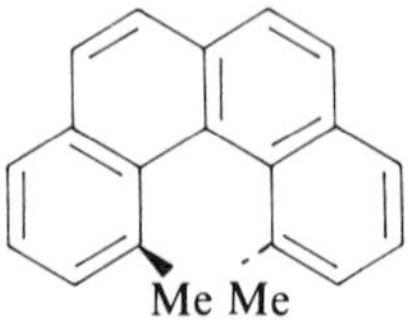

Dimethyl[4]helicene
resolvable

[6]Helicene
resolvable

These helicenes are stable compounds. Helicenes with four or more rings have been resolved, as in the case of [6]helicene and the dimethyl[4]helicene shown above. This is because steric interaction between the terminal rings forces the helicenes to take up helical structures, which may be right- or left-handed.

Partially angular acenes are also known, e.g.

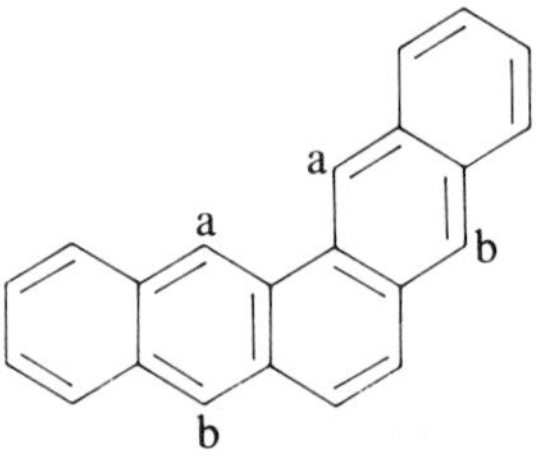

This compound absorbs light at shorter wavelength than does the linear pentacene. It resembles both anthracene and phenanthrene in its properties. It undergoes cyclo-addition reactions across the *meso* positions marked a, b.

Some Other Polycyclic Benzenoid Compounds

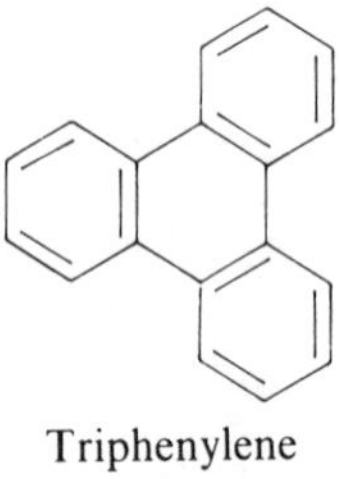

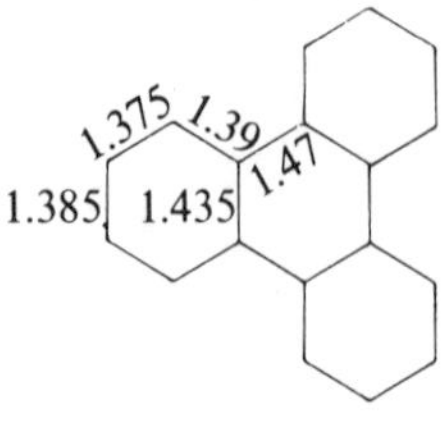

Bond lengths (Å)
in triphenylene

1.375   1.39

1.385   1.435   1.47

Triphenylene

Triphenylene is made up of three benzene rings linked by single bonds.

The π-electrons form three separate six-electron systems in the outer rings, and the central ring can be regarded as a hole between them. It is an extremely stable compound and is neither protonated nor sulphonated in concentrated sulphuric acid, but undergoes other electrophilic substitution reactions such as halogenation, nitration and acylation.

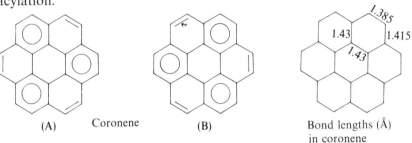

(A)   Coronene   (B)   Bond lengths (Å)
                                      in coronene

Coronene is very stable and has no properties associated with double bonds. It is a hybrid of two equivalent structures (A) and (B), each of which consists of three benzene rings linked by ethylene bridges. The bond lengths are consistent with this hybrid structure.

In contrast, *hexabenzocoronene*, which is also very stable, consists of seven isolated six-π-electron systems linked by single bonds:

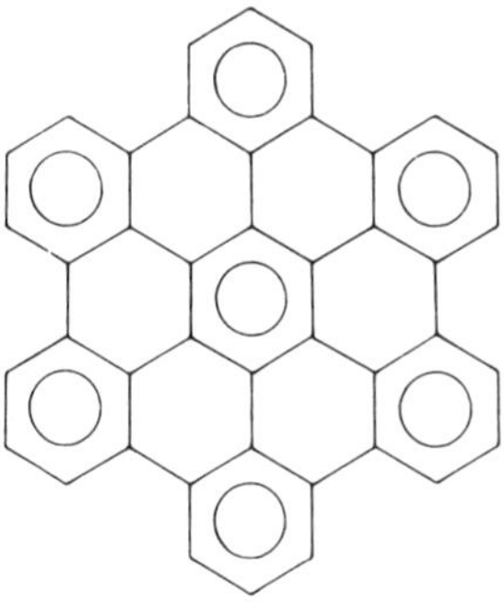

In all polycyclic benzenoid compounds (and indeed non-benzenoid compounds as well) an overwhelming driving force appears to be to attain the maximum number of delocalized systems of six π-electrons.

In concluding this section mention must be made, for different reasons, of the following polycyclic compounds:

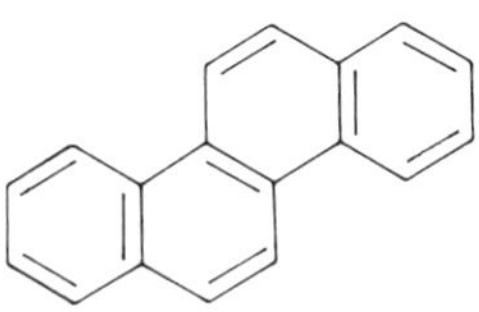

Chrysene 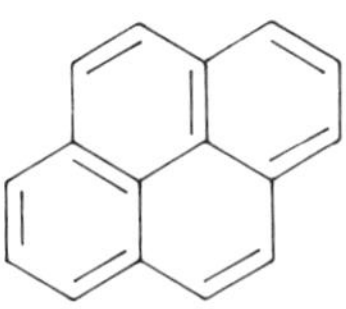   Pyrene 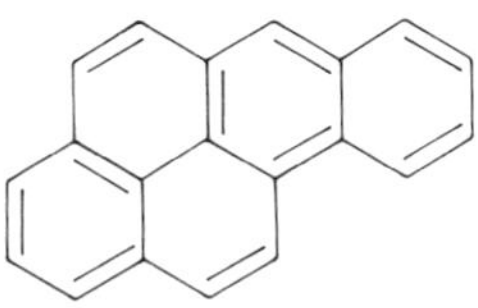   Benz[*a*]pyrene

With a number of other polycyclic compounds they are very potent carcinogens, and exist in soot from coal and in exhaust fumes from cars.

Coal itself is made up of polycyclic hydrocarbons which break down when heated to give naphthalene, anthracene and many other smaller polycyclic compounds, including these carcinogens. The ultimate polybenzenoid compound is graphite, which consists of flat layers of fused six-membered rings, with carbon–carbon bond lengths of 1.42 Å.

Benzoannulenes

A number of benzoannulenes have been prepared. Fusion of benzene rings commonly decreases the delocalization of electrons in annulene rings and lowers their stability.

Some of the first annulenes to be prepared were tetrabenzo-derivatives; it was hoped that the benzene rings would help to stabilize the compounds. Examples are as follows:

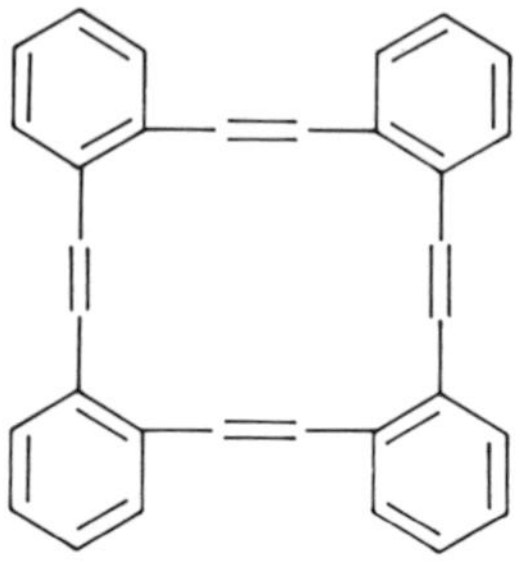

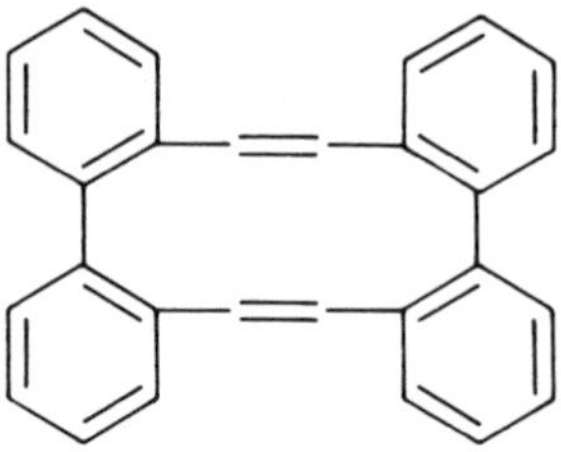

These tetrabenzo[16]- and -[12]annulenes had buckled annulene rings. This removes angle strain in the rings and also inhibits destabilizing interaction of the $4n$-π-electron perimeters. Both compounds are in fact best thought of as systems made up of linked benzene rings.

Other examples of benzoannulenes are:

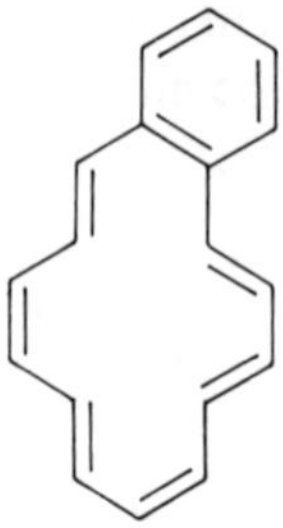

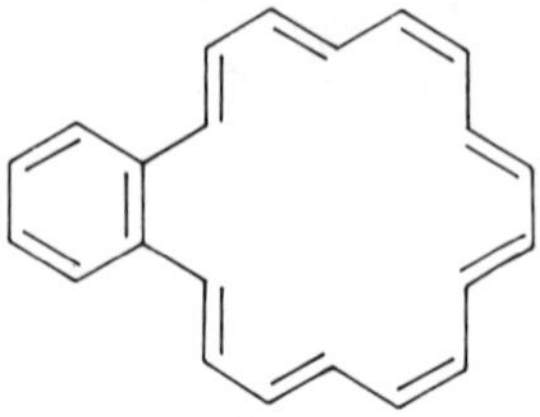

These monobenzo[14]- and -[18]annulenes each have annulene rings which are less diatropic than the corresponding parent annulenes. Thus, whereas [18]annulene provides signals in its ^{1}H-n.m.r. spectrum at $\delta9.28$ and -2.99 for its outer and inner protons, respectively, the benzo derivative provides corresponding signals at $\delta6.45$–8.05 and 4.75–4.95, respectively. Other studies of benzodehydroannulenes and bridged benzoannulenes provide similar findings. Both the diatropicity of $(4n + 2)$ rings and the paratropicity of $(4n)$ rings are lowered by benzoannelation, the effect being more marked in diatropic rings.

An interesting example of this effect of annelation is seen in the case of a dibenzomethano[10]annulene, which appears to exist in a norcaradiene form, with no observable equilibration with the [10]annulene form:

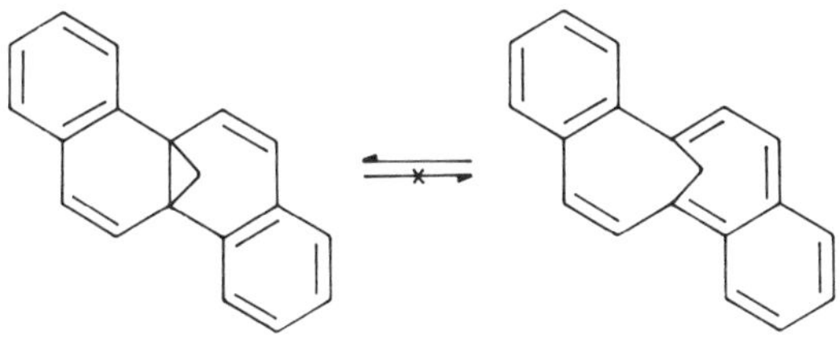

When more than one benzene ring is fused to a larger annulene ring the symmetry of the resultant molecules is an important factor, and the effect on the ring current in the annulene ring depends on the disposition of the benzene rings. If the alternative Kekulé structures of the annelated annulene are identical (as are the two Kekulé structures of benzene) then the annulene ring is fairly strongly diatropic. If the alternative Kekulé structures are not equivalent the diatropicity is greatly reduced.

A converse interaction, whereby a fused ring affects the structure of a benzene ring, is seen in benzocyclobutadienes:

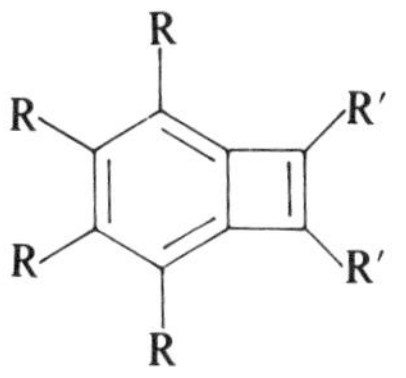

Some fixation of double bonds is induced in the benzene ring in order to lessen the generation of destabilizing cyclobutadienoid character in the four-membered ring. The X-ray structure of a highly substituted derivative ($R = Me, R' = Bu^t$) confirms this prediction. There appears to be some, but diminished, delocalization in the six-membered ring; the single and double bonds in the four-membered ring are, respectively, unusually long and short, the latter despite crowding caused by the t-butyl groups. A similar situation is seen in dibenzocyclobutadiene or biphenylene. (See Chapter 4.)

The tetrabenzo[24]annulene shown below and its [20]-, [28]- and [32]-analogues differ from the benzoannulenes considered hitherto in that more than one edge of the benzene rings are involved in the annulene ring:

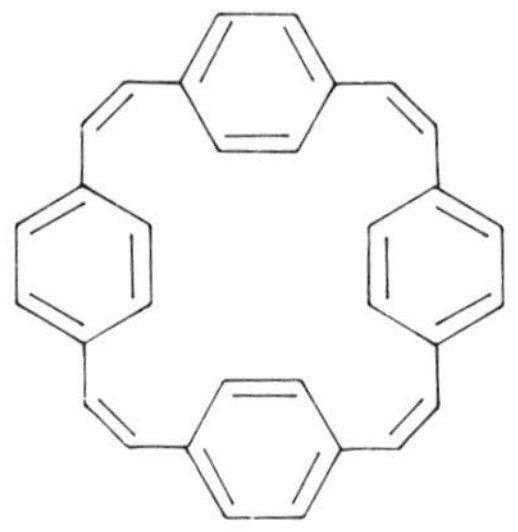

Alternatively, these compounds may be regarded as tetraethylene-bridged annulenes. When they are reduced, first to dianions and then to tetra-anions, the species obtained are, respectively, strongly diatropic and strongly paratropic. The tropicity may be enhanced in these anions because delocalization of charge and of excess electrons is also achieved thereby. This increased contribution of peripheral delocalization on anion formation is also evident from variable-temperature n.m.r. studies, which show that in the neutral molecules spinning of the benzene rings about the adjacent bridging bonds is very rapid, in the dianion it is slow, and in the tetra-anion no rotation is observed and the structure is rigid. These ions are highly conductive.

Benzoannulenide and Benzoannulenium Ions

The stabilization of a cyclopentadienide system or a tropylium system is reduced by annelation of a benzene ring. This is shown by the reduced acidity of indene (benzocyclopentadiene) and fluorene (dibenzocyclo-pentadiene) compared to cyclopentadiene itself, and by the pK_{R^+} values (see Chapter 9) of tropylium salts. The pK_{R^+} values for the tropylium cation and its mono-, di- and tribenzo derivatives are, respectively, 4.75, 1.7, -3.7 and ~ -15. The last figure suggests that there is little, if any, special stabilization in the case of the tribenzotro-pylium cation. This decrease in stabilization in benzo-annelated rings is accompanied by a decrease in delocalization in the annelated ring. In both benzocyclopentadienide and benzocyclononatetraenide anions delocalization may involve the periphery of both rings, possibly in order to delocalize the charge more widely.

Further Reading

For discussion of compounds with fused benzene rings see E. Clar, *The Aromatic Sextet*, John Wiley, Chichester, 1972.
For benzannulenes, see Chapter 7 of the author's 1984 book (for details see Preface).

12

Annulenoannulenes

Bicyclic compounds in which two annulene rings are fused together are called annulenoannulenes. The sizes of the rings are indicated in the usual way by numbers enclosed in square brackets. Thus, if two fused annulene rings are each [14]annulene rings, the bicyclic compound is called [14]annuleno[14]annulene. The benzo derivatives considered in Chapter 11 are special cases of annulenoannulenes; naphthalene is [6]annuleno[6]annulene. When smaller rings are involved the normal bicyclo nomenclature is commonly used. Thus [4]annuleno[8]annulene is usually described as bicyclo[6.2.0]deca-pentaene. All the annulenoannulenes with larger rings which have been reported hitherto are in fact dehydroannulenes.

Calculations suggest that the electronic structure and properties of annulenoannulenes are determined more by the nature of the individual fused rings than by the size of the overall perimeter, and experimental work generally concurs with this.

An example of an annulenoannulene having one bond common to each ring is the following tetrakisdehydro[14]annuleno[14]annulene:

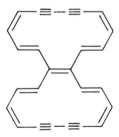

The ring currents in each ring are smaller than that in the related bisdehydro[14]annulene, but not as small as that in the [14]ring of the corresponding benzobisdehydro[14]annulene.

[14][16]- and [14][18]-analogues of this compound have been prepared and it is found that in these compounds the diatropicity of

the [14]ring decreases as the size of the fused ring increases, irrespective of whether this fused ring is a [4n] or [4n + 2]ring.

A number of annulenoannulenes have been prepared with three (and also with five) bonds common to the two rings. An example is the [14]annuleno[14]annulene shown:

An X-ray study (at − 150 °C) and ^{13}C-n.m.r. spectra indicate that the π-electrons are delocalized as in two [14]rings, rather than as a bridged peripheral 22-π-electron system. Similar results have been obtained with other annulenoannulenes of this kind.

The examples just given each have two rings of the same size. If two [4n + 2] rings of different size are fused together, the ring which is inherently more diatropic, i.e. the smaller one, tends to suppress the diatropicity in the other ring. This has already been noted in Chapter 11 for the benzoannulenes. If a [4n] and a [4n + 2] ring are fused together, the paratropic ring current in the [4n] ring is largely retained, but there is partial suppression of the diatropic ring current in the [4n + 2] ring. These effects have been shown to take place in a variety of annulenoannulenes.

Some of the smaller ring annulenoannulenes are worth individual discussion in order to demonstrate the interplay of electronic and steric factors which are involved. The [4][8] and [8][8] compounds will be considered.

Bicyclo[6.2.0]decapentaene

A number of conflicting factors come into play in determining the detailed structure of this molecule. If the bridging bond is completely single, and if the molecule adopts a planar conformation, then a delocalized system of ten π-electrons can be established in the periphery. If, however, it adopts a planar conformation, then there is more strain in the molecule than there is in a conformation in which the eight-membered ring is buckled. These two effects oppose one another. In addition, peripheral delocalization involves a measure of cyclobutadienoid character in the four-membered ring, arising from the alternative Kekulé form to that shown above, unless the bridging bond plays no part whatsoever in π-interaction. Without this last proviso, the setting up of a stabilizing peripheral ten-π-electron system would be accompanied by some setting up of a destabilizing four-π-electron system in the four-membered ring. It can be expected that there is a subtle balance between these different electronic and steric factors. Calculations suggest that this molecule should exist as a peripheral ten π-electron system in which little delocalization of electrons takes place, and that the bridging bond acts merely as a cross-ring σ-bond.

The compound itself is a red-orange oil which is sensitive to air. A number of derivatives with substituents in the four-membered ring have also been prepared. An X-ray crystal structure of a diphenyl derivative shows that the molecule is almost planar, and that the bridging bond is very long for a bond linking two sp^2 carbon atoms (1.53–1.54 Å). There is some small alternation in the peripheral bond lengths, but some delocalization of π-electrons is indicated; the delocalization is less than in 1,6-methano[10]annulene. However, it is enough to provide delocalization energy sufficient to compensate for the strain energy involved in the planar conformation.

However, the bicyclo[6.2.0]decapentaenes have sufficient alkene character to undergo cyclo-addition reactions, e.g.

Octalene or Bicyclo[6.6.0]tetradecaheptaene

(A)   (B)   (C)

This compound is usually known by its trivial name, *octalene*. It has been prepared starting from tetrahydronaphthalene, by successive ring-expansion steps. Possible structures to be considered are the localized forms (A) and (B) and the delocalized structure (C). Structure (C) would result in lower energy because of the peripheral 14-π-electron system, but, conversely, energy would be required to flatten the molecule, which results in introduction of strain into the system.

The spectra of octalene leave no doubt that the molecule is non-planar and exists predominantly in structure (A). Variable-temperature n.m.r. studies show that both conformational and valence isomerism processes take place, the latter only at higher temperatures. It thus appears that the energy required to flatten the structure must be greater than that released in the setting up of the delocalized π-electron system.

Reaction of octalene with lithium provides in turn a radical anion, a dianion, a radical trianion and a tetra-anion. The n.m.r. spectra of the dianion suggest that it is best represented as an equilibrating system of valence isomers:

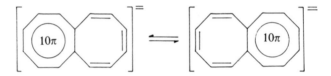

This is a bridged delocalised ten-π-electron system, which is preferred to a peripheral system of 16 π-electrons, since the latter is a destabilizing [4n] system. The tetra-anion is diatropic and apparently has a peripheral delocalized system of 18 π-electrons. It is stable for several days in solution in tetrahydrofuran so long as air is excluded.

Further Reading

For a review on annulenoannulenes see M. Nakagawa, *Angew. Chem.*, **91**, 215 (1979); *Angew. Chem. Int. Edn Engl.*, **18**, 202 (1979).

13

Azulenes, Pentalenes, Heptalenes and Related Compounds

The annulenoannulenes which were considered in Chapter 12 were made up of two fused rings, each of which had an even number of atoms. For these compounds it is possible to draw structures made up of alternate single and double bonds in such a way that the bond common to both rings is either a single or a double bond. This is typified by either naphthalene or octalene:

For fused-ring compounds with odd numbers of atoms in the rings, such as pentalene, azulene or heptalene, it is not possible to draw structures with alternate single and double bonds unless the bond common to both rings is a single bond:

Pentalene	Azulene	Heptalene

The most common and longest-known of all non-benzenoid fused-ring polyalkene systems is azulene, and for this reason azulenes will be considered first.

Azulenes

Azulenes are by far the most numerous and most investigated of all polycyclic non-benzenoid polyalkenes. The parent compound, *azulene*, bicyclo[5.3.0]decapentaene, can, like benzene, be drawn in two equivalent Kekulé forms. It has a peripheral conjugated system involving ten π-electrons and has an intense blue colour; its name is derived from this property (cf. azure). This colour is perceptible at dilutions as great as 1 in 10 000 in a solvent such as light petroleum.

The development of blue colours in certain essential oils (e.g. oil of camomile) was noted as far back as the fifteenth century. Many of these oils contain sesquiterpenes which are hydrogenated azulene derivatives, and operations such as distillation, steam distillation or· aerobic oxidation give rise to the formation of azulenes. For example, an intensely blue-coloured fraction was obtained from the distillation of geranium oil. These blue colours were caused by the formation of azulene derivatives. Being derived from sesquiterpenes, many naturally derived azulenes have the molecular formula $C_{15}H_{18}$, and are dimethylisopropylazulenes.

An important step in the elucidation of the structure of azulenes was the discovery that azulenes can be extracted into aqueous mineral acids, in which they give pale-brown solutions, and can be recovered unchanged from these solutions by diluting them with water. This made it possible to isolate azulenes from the other organic material which accompanied them when they were obtained from natural sources.

Synthesis of Azulenes

Azulene was in fact prepared, unknowingly, in 1893, as a by-product in the preparation of cyclopentanone by the distillation of calcium adipate. All the earlier methods used to prepare azulene and its derivatives involved initial synthesis of bicyclo[5.3.0]decane derivatives which then had to be dehydrogenated at a high temperature, and great losses were often encountered at this stage.

More recent methods remove the necessity of this dehydrogenation step. Most involve elegant adaptations of an alkali catalysed condensation of glutacondialdehyde with cyclopentadiene. This may be represented as:

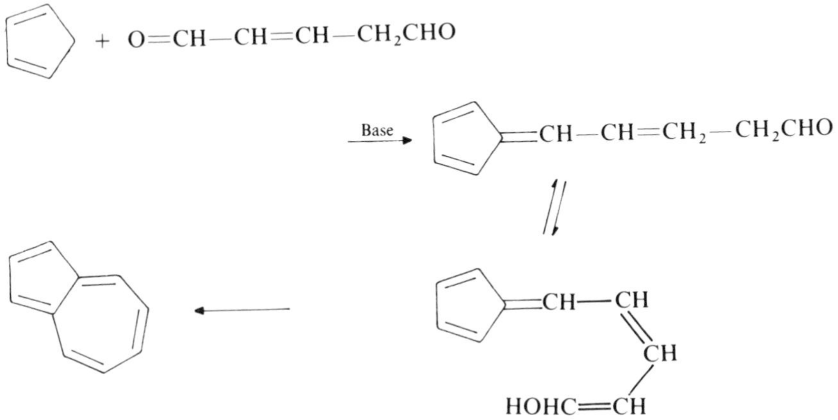

Glutacondialdehyde itself is not easy to handle, but various derivatives of it are used instead. For example, treatment of pyridine with 2,4-dinitrochlorobenzene, followed by *N*-methylaniline, provides a salt of a dianil derivative, and this derivative, on reaction with cyclopentadiene in the presence of alkali, provides azulene:

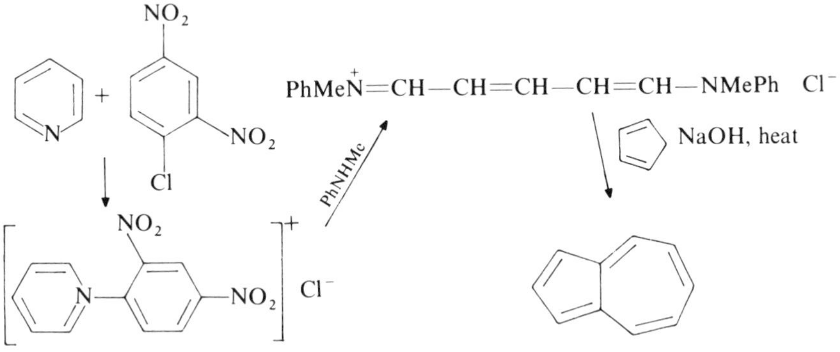

A simple modification of this reaction involves the treatment of an *N*-alkylpyridinium salt with sodium cyclopentadienide:

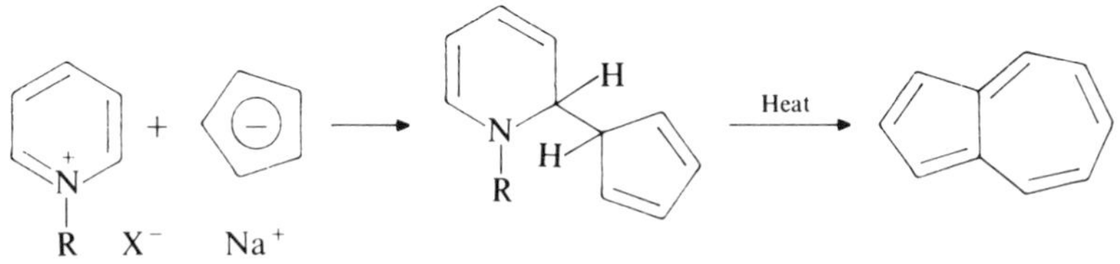

A substituted pyrylium salt may take the place of the pyridinium salt, and in this case reaction occurs at room temperature. The reaction is not successful, however, starting from unsubstituted pyrylium perchlorate:

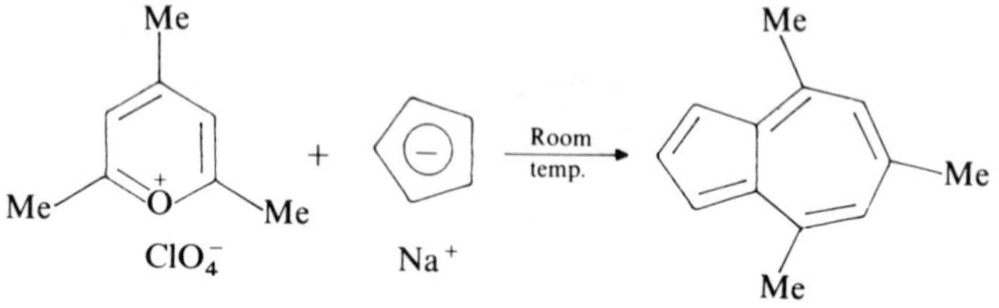

By these and a number of other different methods a wide range of azulene derivatives has been synthesized.

Structure of Azulene

The ^{1}H-n.m.r. spectrum of azulene, with signals in the range $\delta 6.92$–8.12, indicates that it is diatropic. In addition to the two Kekulé-type structures (A, B), it is possible to draw a number of dipolar structures (C–G):

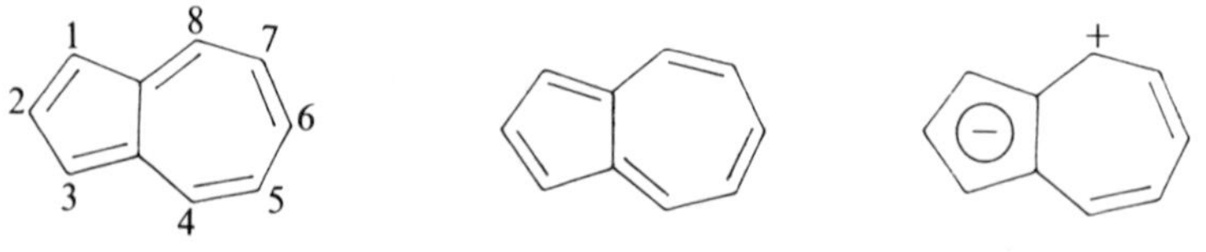

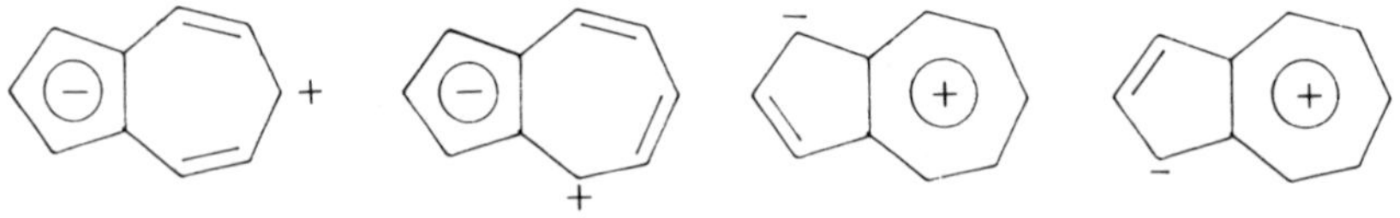

In the dipolar structures (C), (D), (E) the five-membered ring has six electrons and a negative charge delocalized as in a cyclopentadienide anion, and a positive charge appears at the 4-, 6- and 8-positions. In structures (F) and (G) the seven-membered ring resembles a tropylium ion and a negative charge appears in the five-membered ring at the 1- and 3-positions. Note that, as in the case of naphthalene, it is not possible to have delocalized sextets of electrons simultaneously in both rings; only ten π-electrons are available. Azulene has only a very small dipole moment (0.80D), so that any contribution from the dipolar forms is very small, but it is none the less real. Azulene is best regarded as a hydrocarbon having a cyclic conjugated system of ten π-electrons, but this electron system is not evenly dispersed over the molecule, due to the small contribution from these dipolar forms.

There has been much theoretical investigation of the extent of bond localization or delocalization in the periphery of azulene. It appears that there is only a small difference in energy between the localized and delocalized structures, with the latter being marginally more stable. X-ray data both on azulene and on derivatives are inconclusive; the crystal structure of azulene is disordered. Because of the small difference in energy between localized and delocalized forms it is possible that the disorder in the crystals reflects valence isomerism between localized and delocalized forms rather than, or in addition to, disordered packing of the molecules in the crystal. All calculations and X-ray studies agree that the bridging bond is long (measured as 1.482 Å), and plays little part in any conjugation in the molecule, although, as will be seen, it plays an important role in the chemistry of azulenes.

Reactions with Electrophiles and Nucleophiles

The solubility of azulenes in strong acids is due to formation of azulenium ions:

The electronic structure of azulene is disrupted, resulting in loss of the characteristic blue colour, and the product contains a tropylium ion.

Other electrophiles react similarly with azulene to give intermediates which can lose a proton to provide a substituted azulene:

E^+ = electrophile

This intermediate is stabilized by delocalization of the positive charge over the seven-membered ring in a tropylium ion, and its formation is thereby favoured. Reaction takes place at the 1- or 3-positions.

Similarly, nucleophilic attack on the seven-membered ring, which takes place at the 4-, 6- or 8-positions, produces an intermediate in which the negative charge is delocalized over the five-membered ring:

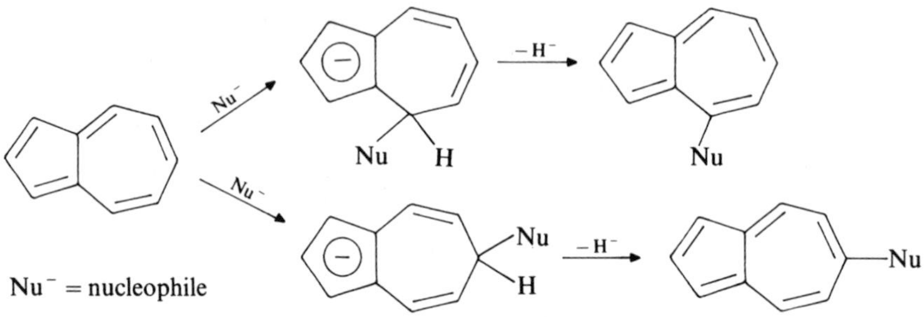

Nu⁻ = nucleophile

Thus attack by an electrophilic or nucleophilic reagent on the appropriate ring of azulene results in the formation of a species having a delocalized sextet of electrons. In these reactions, rather than a stabilized delocalized system of electrons being lost, one such system (the peripheral) is exchanged for another (a tropylium ion or a cyclopentadienide ion). The transition states leading to these intermediates are accordingly lowered in energy, and electrophilic or nucleophilic attack is thereby facilitated.

The role of the bond common to both rings in these reactions may be noted. If no such bond were present, intermediates involving such cyclic stabilized ions would not be possible, i.e.

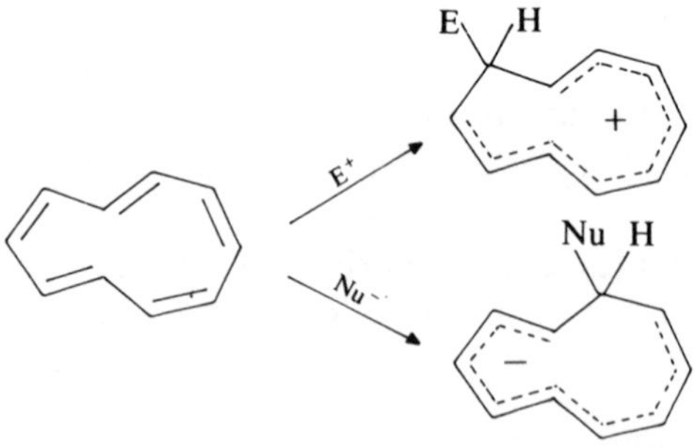

The fact that electrophilic or nucleophilic attack results in substitution reactions indicates, however, that regeneration of the azulene system is energetically favoured, i.e. it is itself a stabilized system.

Azulenium salts have been isolated as crystalline products by the action of perchloric acid or tetrafluoroboric acid in ether on azulene derivatives. These salts are colourless or pale brown. Addition of water immediately restores the blue colour of the parent azulene.

Azulenes undergo electrophilic substitution very readily; their reactivity is comparable to that of a reactive benzene derivative such as aniline. Thus, for example, azulenes can be halogenated, nitrated, sulphonated and acylated and couple with diazonium salts. Depending on the conditions used, 1-substituted or 1,3-disubstituted products are obtained. If the 1- and 3-positions are blocked by substituents, substitution by electrophiles may take place at the 5-position. An example is the reaction of 1,3-dichloroazulene with acetyl chloride and tin(IV) chloride to give 5-acetyl-1,3-dichloroazulene.

Nucleophilic substitution of azulenes occurs preferentially at the 4- or 8-positions, or, if these positions are already occupied by substituents, at the 6-position. However, study of these reactions has been hampered by the relative lack of stability of azulenes towards bases, presumably resulting from the high reactivity of the intermediate cyclopentadienide system. Azulene is rapidly decomposed when it is heated with alcoholic alkali. With lithium alkyls azulenes give alkylazulenide anions, which react with water to give dihydroazulenes. Alkylazulenes may be prepared in this way:

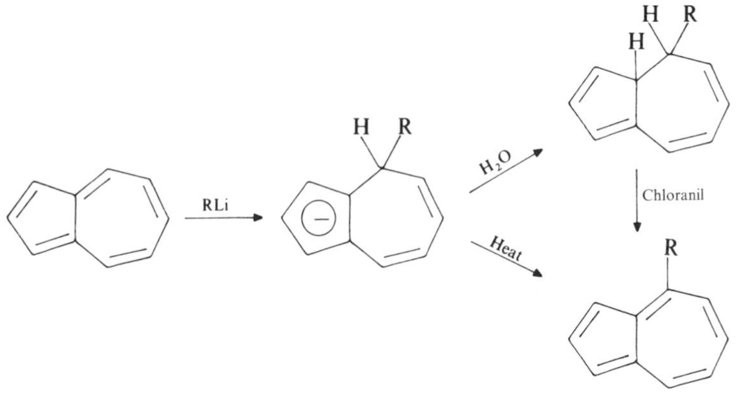

Other Reactions of Azulenes

Most dienophiles do not take part in cyclo-addition reactions with azulenes but rather react by conjugate addition. Thus maleic anhydride gives an azulenylsuccinic anhydride:

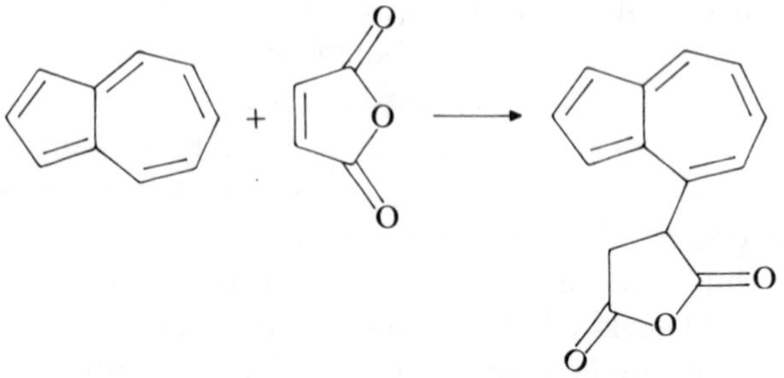

However, alkynes may undergo cyclo-addition across the 1- and 4-positions of azulene. Thus benzyne reacts as shown:

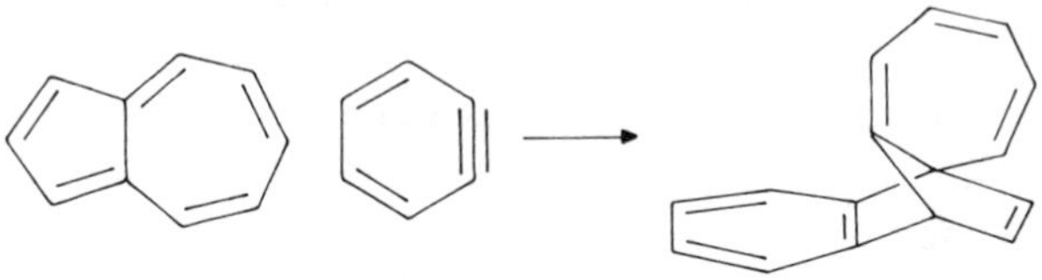

Dimethyl acetylenedicarboxylate reacts similarly and the product rearranges to give a heptalene derivative; this is a convenient means of access to heptalene compounds:

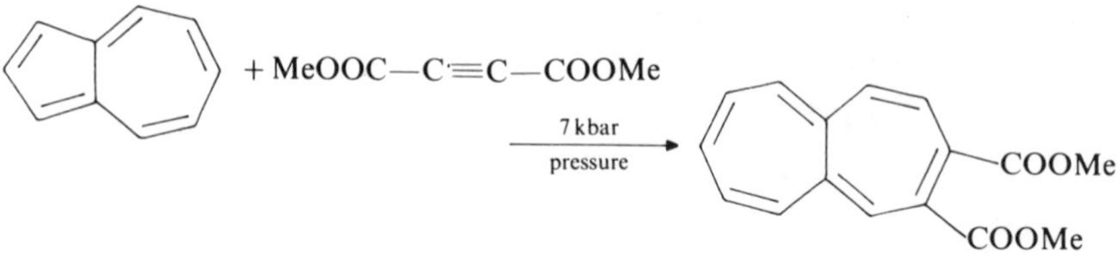

Azulenes are susceptible to oxidation and decompose slowly on standing; this is hastened by light. They are readily degraded by permanganate.

Reduction of azulene by sodium in liquid ammonia gives 1,6-dihydroazulene. Azulenes are readily reduced by sodium and ethanol, or catalytically, to give polyhydroazulenes. It is possible, however, to reduce unsaturated substituent groups such as alkene or carbonyl groups without reducing the azulene ring.

Homoazulene

1,5-Methano[10]annulene, mentioned in Chapter 5, has also been given the name *homoazulene*, since it has a structural resemblance to azulene, with a bridging methylene group replacing the bridging single bond of azulene:

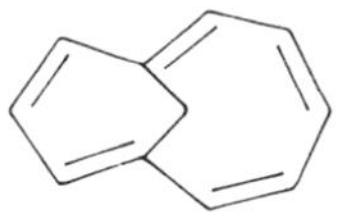

This compound is orange, and its electronic spectrum differs from that of 1,5-methano[10]annulene similarly to the way in which the electronic spectrum of azulene differs from the spectrum of naphthalene.

Pentalenes

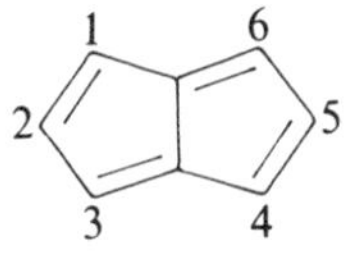

The periphery of pentalene has eight π-electrons and thus would not be expected to provide special stabilization. The only pentalenes to have been isolated as stable compounds are either substituted by large groups which sterically protect the molecule (e.g. 1,3,5-tri-t-butylpentalene) or by 1,3-amino substituents.

Attempted preparations of pentalene itself and of methyl derivatives have provided dimers. These dimers have been decomposed photolytically at low temperatures to generate pentalenes which dimerize again when the temperature is raised.

1,3,5-Tri-t-butylpentalene forms deep-blue crystals which are stable for several hours at room temperature in the absence of air. The double bonds are localized although there appears to be valence isomerism between the two Kekulé forms. N.m.r. spectra indicate that the compound is paratropic.

1,3-Bis(dimethylamino)pentalene forms dark-blue crystals which are stable towards oxygen for several hours at room temperature:

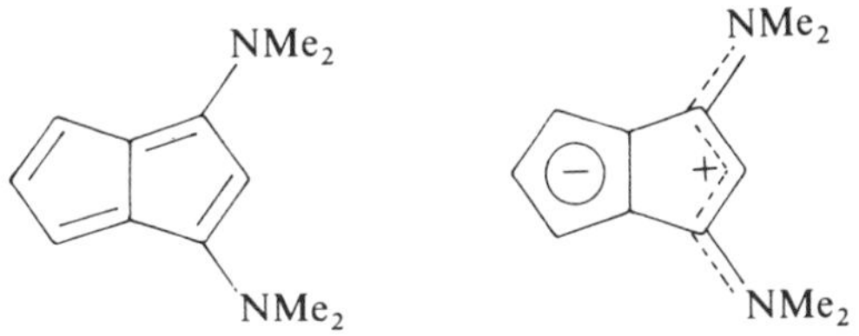

This compound may owe its stability to the possibility of its adopting the electronic structure of a vinamidinium cyclopentadienide as shown in (A). In accord with this, it is protonated by acid at the 2-position, a reaction typical of vinamidinium systems.

Pentalenide Salts

Dilithium pentalenide has been prepared by treatment of 1,5-dihydropentalene with n-butyl lithium:

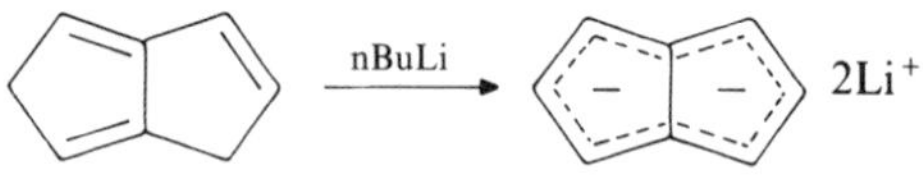

This salt forms colourless crystals and is stable in solution in dry tetrahydrofuran at room temperature. As the dianion of pentalene it has ten π-electrons and this should provide stability. Its ^{1}H-n.m.r. spectrum shows signals at δ4.98 (4H) and 5.72 (2H) and is in accord with the offsetting of the upfield shift due to the double negative charge by a diamagnetic ring current due to the ten-π-electron system. This ion may be regarded as a bridged cyclo-octatetraenide dianion.

Heptalenes

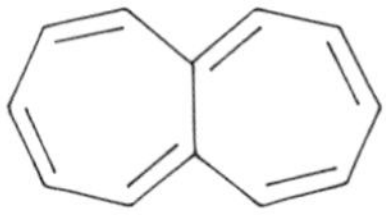

Heptalenes have been made by cyclo-addition reactions of a variety of azulenes with an alkyne and by ring expansion of tetrahydronaphthalene and dehydrogenation of the resultant dihydroheptalene. Like pentalene, heptalene has a peripheral conjugated system containing $4n$ π-electrons, so neither special stabilization of the molecule nor delocalization of the π-electrons should be expected. Unlike pentalene, heptalene has the option of taking up a non-planar shape. Spectroscopic evidence indicates alternate single and double bonds but also fast valence isomerism between the two Kekulé forms. The ^{1}H-n.m.r. spectrum is consistent with either a small paramagnetic ring current or a lack of ring current; the latter could result from lack of planarity in the molecule, which would reduce strain.

Increasing substitution by methyl groups rather surprisingly leads to increased stability, and some methylheptalenes, unlike heptalene itself, are both thermally stable and stable to air. It has been suggested that an important factor here is increased twisting of the perimeter caused mainly by steric crowding due to increasing numbers of methyl groups. This reduces the cyclic conjugation and increases the HOMO–LUMO gap of the system.

Heptalene is readily protonated in strong acids to give a cycloheptatrienotropylium salt:

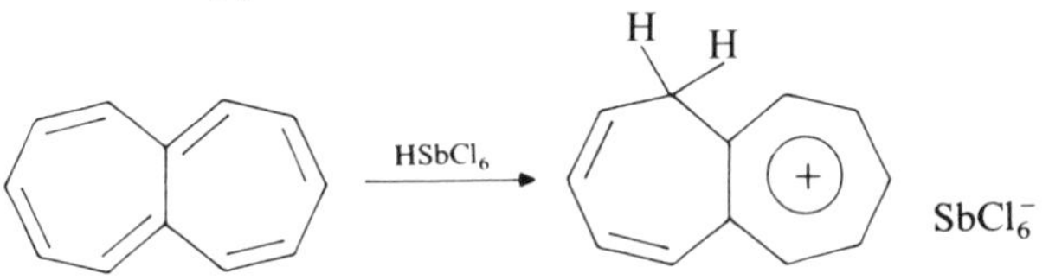

Treatment of heptalene with lithium in tetrahydrofuran results in the formation of a dianion, whose ^{1}H-n.m.r. spectrum indicates that it is diatropic, as would be expected for a species having $(4n + 2)$ π-electrons. This dianion is thermally stable, much more so than heptalene itself.

Polycyclic Compounds Related to Azulene

A number of polycyclic compounds made up from rings with odd numbers of atoms are known. Their detailed electronic structures are complicated. A simple example is the following cyclopentazulene:

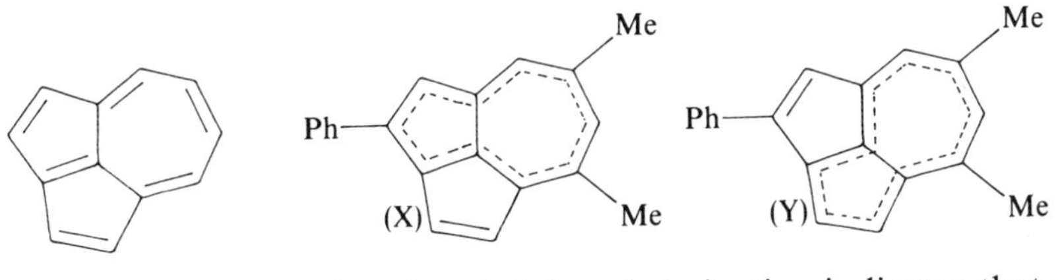

An X-ray structure of a dimethylphenyl derivative indicates that it exists as a hybrid of the two structures (X) and (Y). Both (X) and (Y) represent etheno-bridged azulenes. There is obviously an energetic drive to provide a ten-π-electron azulene-type system, but this is shared between the two five-membered rings.

An X-ray crystallographic examination of the azulenoazulene shown below was best interpreted as indicating a delocalized azulene-like portion and a bridging alkene portion:

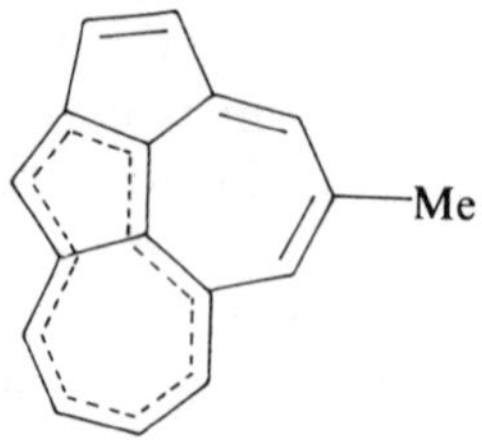

It is, however, difficult to generalize about the structures of this type of polycyclic compound, but the tendency does seem to be towards setting up, wherever possible, of $(4n + 2)$-π-electron systems within the structures, or, alternatively, the avoidance of circuits of $4n$-π-electrons.

The tricyclic compound shown below can be considered as a homologue of azulene. It has a doubly bridged 14 π-electron periphery:

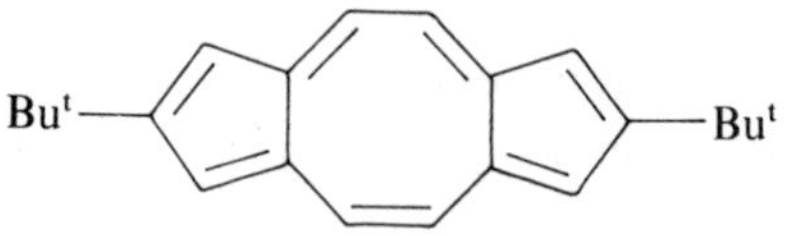

It is a stable red-brown compound. An X-ray crystal structure indicates that all the peripheral bond lengths are very similar, at $c.$ 1.40 Å, whereas the bridging bonds are long (1.49 Å). The ring system is planar. NMR spectra are also consistent with a delocalized π-electron system in the periphery, and the electronic spectrum resembles that of azulene, although shifted to longer wavelengths.

Further Reading

Articles in *Methoden der Organischen Chemie* (*Houben–Weyl*), 4th edition, Vol. 5/2c, Thieme, Stuttgart, on azulenes: K.-P. Zeller, pp. 127ff.; on pentalenes: H. J. Lindner, pp. 103ff.; on heptalenes: G. Becker and H. Kolshorn, pp. 418ff.

Azulenes: T. Nozoe, *Pure Appl. Chem.*, **28**, 239 (1971); V. B. Mochalin and Y. N. Porshnev, *Russ. Chem. Rev.* (*Engl.*), **46**, 530 (1977).

Pentalenes and heptalenes: K. Hafner, *Pure Appl. Chem.*, **28**, 153 (1971).

General comments on polycyclics: C. Glidewell and D. Lloyd, *Tetrahedron*, **40**, 4455 (1984); *J. Chem. Ed.*, **63**, 306 (1986); on polycyclics with odd-numbered rings: *Acta Script.*, **28**, 385 (1988).

See Chapter 8 of the author's 1984 book (for detail see Preface).

14

Aromatic Transition States

This book is primarily concerned with compounds, of varying degree of stability, but all identifiable as entities. In 1939 M. G. Evans and E. Warhurst (*Trans. Farad. Soc.*, **34**, 614 (1938); **35**, 824 (1939)) put forward the concept that the same stabilization of cyclic compounds which derives from circuits of $(4n + 2)$ electrons might also apply to cyclic transition states, drawing a comparison between the electronic structure of benzene and the transition state of a Diels–Alder reaction. Both involve cyclic interactions between six π-electrons. This may be expressed graphically as follows:

Later the whole pattern of *pericyclic reactions*, i.e. ring-forming reactions involving the cyclization of one or more unsaturated compounds by a one-step concerted pathway, and the converse ring-opening reactions, was rationalized in the Woodward–Hoffmann Rules [see *Angew. Chem. Int. Edn Engl.*, **8**, 781 (1969); *Angew. Chem.*, **81**, 797 (1969)]. A number of different theoretical approaches to these reactions and rules have been used but will not be considered here. Rather, in view of the overall subject matter of this text, some brief mention of the concept of aromatic transition states is added, as an addendum to the main text. An authoritative review, by M. J. S. Dewar, is available (*Angew. Chem. Int. Edn Engl.*, **10**, 761 (1971); *Angew. Chem.*, **83**, 859 (1971)).

Evans and Warhurst pointed out that not only might the transition state of the Diels–Alder reaction be stabilized by the involvement of six π-electrons but also that no such stabilization could be expected for a cyclic dimerization reaction of ethylene leading to the formation

of cyclobutane, since this would involve interaction between four
π-electrons, not favoured by analogy with Hückel's rule.

Similarly, two butadienes do not undergo cyclo-addition reactions
involving eight π-electrons to give eight-membered rings, but rather
take part in [4 + 2] cyclo-addition reactions:

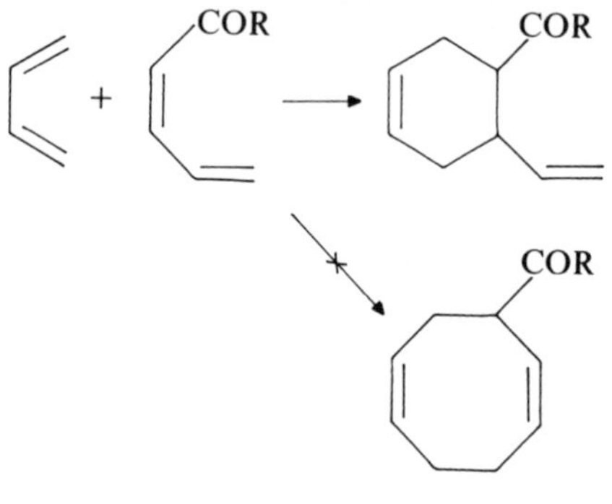

The obvious extension of this is that any such cyclo-addition
reaction involving (4n + 2) electrons in its transition state might be
energetically favoured, because of stabilization of this transition state.
As is the case when compounds are considered, so with transition
states other relevant factors, for example, steric factors, must also be
taken into account in assessing the situation.

Other Cyclo-addition Reactions

In the Diels–Alder reaction one reactant provides four π-electrons
and the other provides two π-electrons, so this may be called a [4 + 2]
cyclo-addition reaction. An example of a [6 + 4] cyclo-addition
reaction, providing a ten-π-electron transition state, was mentioned in
Chapter 8, namely a reaction of tropone with butadiene:

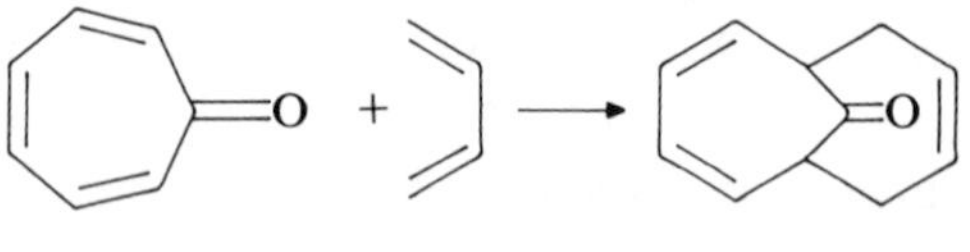

Another ten-π-electron transition state is seen in the [8 + 2] cyclo-
addition reaction of heptafulvene with dimethyl acetylenedicarboxylate:

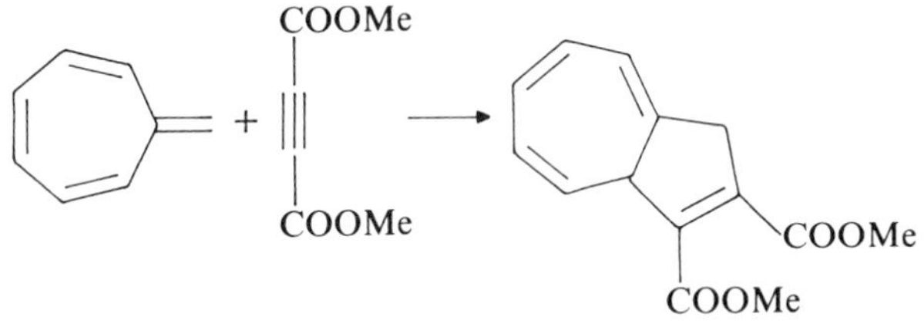

The geometry of the reactants is favourable to the formation of a ten-π-electron transition state in both these examples.

In Chapter 10 a $[14 + 2]$ cyclo-addition reaction of heptafulvalene with tetracyanoethylene was mentioned. This involves a 16-π-electron transition state, i.e. a 4n-π-electron system. However, the stereochemistry of the product shows that the alkene has added in a *trans* mode to the heptafulvalene:

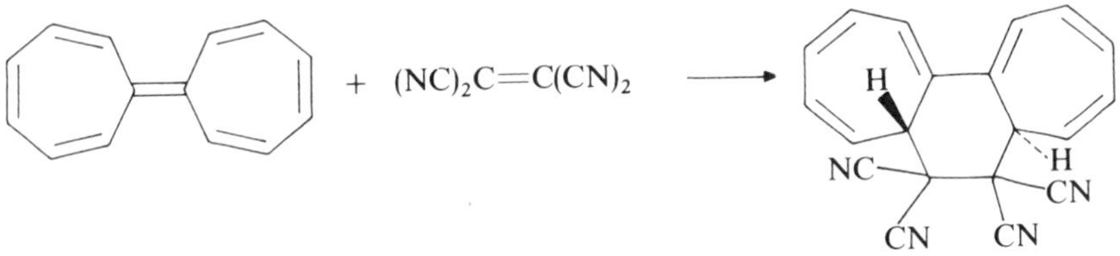

To achieve this geometry a twisted, or Möbius-type, cyclic arrangement of the 16 participating π-electrons must be involved. At the end of Chapter 6 mention was made of the proposal that, for Möbius-shaped annulenes, the predictions of Hückel's rule for planar annulenes should be reversed, and that $[4n]$ Möbius annulenes should be stabilized. This $[14 + 2]$ cyclo-addition reaction provides an example of such a $[4n]$ stabilized transition state. The shape of the reactants enables this $[4n]$ Möbius transition state to be formed. Möbius-type reactions do not provide smaller rings (for example, cyclobutane from ethylene) because geometry prevents the achievement of the necessary twisted transition state.

Electrocyclic Reactions

Other examples of pericyclic reactions are *electrocyclic* reactions, namely reactions involving ring closure of fully conjugated linear polyenes and the reverse ring-opening reactions. A major feature of these reactions is their stereospecificity, as the following examples show:

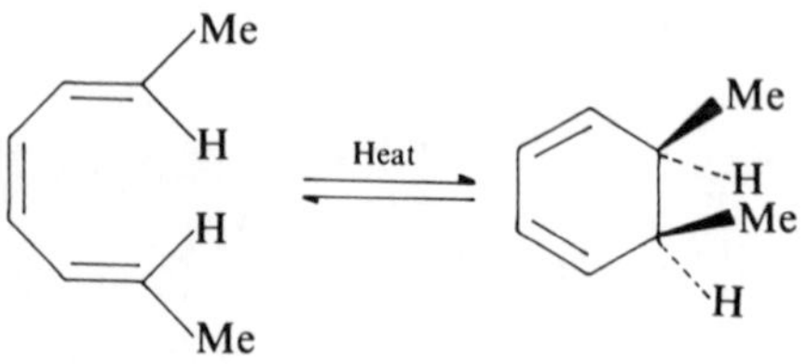

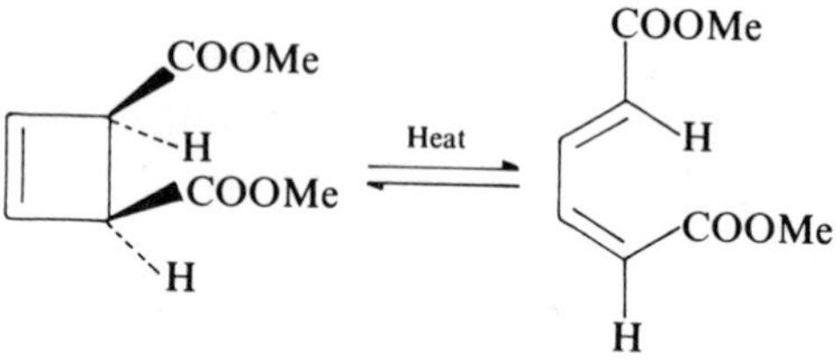

In the first example the two terminal groups rotate in the opposite direction, one clockwise and the other anti-clockwise (or widdershins), as illustrated:

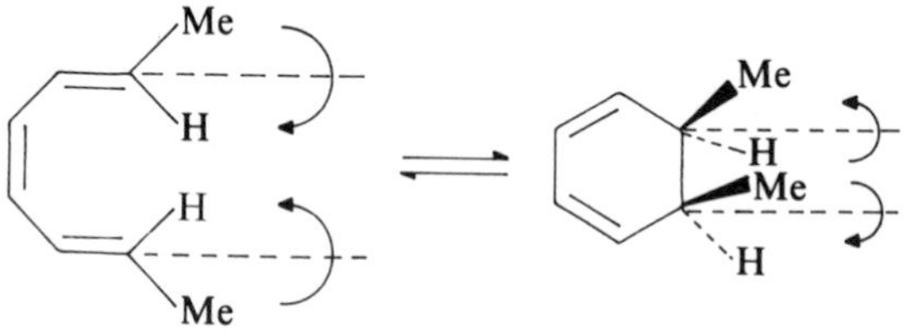

This behaviour is described as *disrotatory*.

In the second example both terminal groups rotate in the same direction:

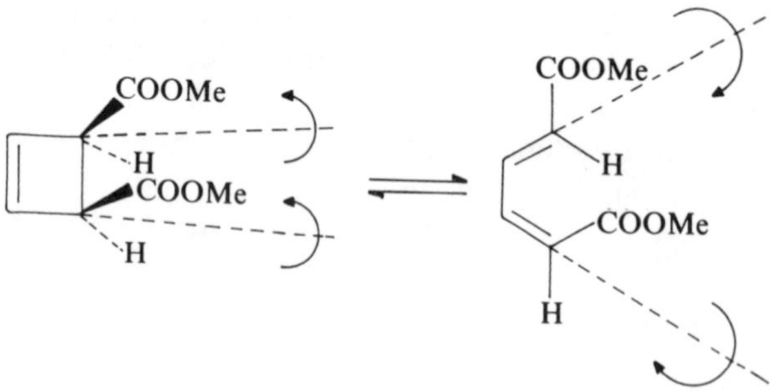

This behaviour is called *conrotatory*.

The identity of the groups involved in the rotation is not a factor; the number of electrons in the transition state is. The first example involves the interaction of six electrons, the second the interaction of four:

This difference in behaviour between systems having, respectively, six and four electrons immediately bears comparison with Hückel's rule, whereby in the case of planar annulenes six π-electrons provide special stabilization, whereas for systems having $4n$ π-electrons stabilization is expected to be associated with Möbius-shaped systems.

Consider the interactions required between the π-orbitals of the open-chain form to create the σ-bond which closes the ring, and to provide the observed stereochemistry (see the diagram on p. 178).

This oversimple pictorial representation shows that in the case of the six-electron system a planar interaction provides ring closure with the observed disrotatory mechanism, while the conrotatory mechanism in the case of the four-electron system involves a Möbius-type interaction. The same argument may be applied to the converse ring-opening reactions.

In general, electrocyclic reactions involving $(4n + 2)$ electrons follow a disrotatory mode, providing a stabilized planar transition state, and those involving $4n$ electrons follows a conrotatory mode via a stabilized Möbius transition state. Once again, so-called aromatic transition states are favoured and determine the stereochemical courses of the reactions. (In photochemically induced electrocyclic reactions these rules may be reversed, because excited states are involved, but this will not be pursued further here; further discussion is readily available in appropriate textbooks.)

Cyclo-addition and electrocyclic reactions are only two examples of pericyclic reactions involving stabilized aromatic transition states.

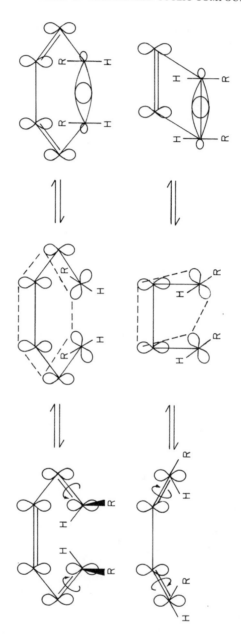

Among others are dipolar cyclo-addition reactions and rearrangements such as Cope rearrangements and hydrogen transfer reactions, e.g.

As stated at the beginning of this chapter, this book is primarily concerned with compounds, but it seemed appropriate to add this very brief mention of the idea that factors considered in the earlier main chapters were not confined in their application to the chemistry of compounds but also influence the transition states involved in chemical reactions.

Kekulé, Robinson and Hückel, to mention but a few—there are many others, famous and not so famous—certainly started something. The subject continues to grow and have applications unthought of by the early workers—and by all of us—for many new aspects will appear and are just waiting for you or me to notice them.

Further Reading

A number of excellent texts are available; for example, T. L. Gilchrist and R. C. Storr, *Organic Reactions and Orbital Symmetry*, Cambridge University Press, Cambridge, 1972 and later editions.

Index